Danforth

Praise for *Born to Believe*

D1023704

"Fascinating, mind-bending reading. . . . Heady stuff, but with extensive research and credible scientific resources to support it, enough to make a person rethink concepts of truth, reality, and belief. So rich a book that it begs to be read in small bites over a long time."

—*Booklist*

"Newberg's neutrality is as scrupulous as a nun's conscience, the neuroscientist and the religious studies professor seemingly in perfect balance."

—Brian Bethune, *Maclean's*

"The book offers a helpful analysis on how to guard against prejudicial thinking and how to distinguish constructive from destructive beliefs as one seeks to be a 'better believer.' This work joins other studies that seek to understand the link between biology and religion."

—*Choice: Current Reviews for Academic Libraries*

"Here is a book that seeks not to dismiss or ignore our will to believe, but instead explores why believing—even secular beliefs—is such a necessary and 'hardwired' aspect of being human. Newberg and Waldman bring an immense scientific learning to this compelling work of immense clarity. *Born to Believe* is certainly the best scientific statement yet on the will to believe."

—Stephen G. Post, PhD, Professor of Bioethics, School of Medicine, Case Western Reserve University, author of *Unlimited Love* and President, Institute for Research on Unlimited Love, Altruism, Compassion, Service

"Why believe what you read—or hear, or think? This intriguing book offers insights into how we can constructively question our beliefs in a way that expands our minds with deeper insights into others, and ourselves. Offering a wide-ranging discussion of beliefs—from the insights gleaned from brain studies of transcendent experiences to explorations of perceptual distortions—the authors walk us through an adventure in thinking that is sure to raise as many questions as it answers in its illuminating discussions."

—Daniel J. Siegel, MD, author of *Mindsight, Our Seventh Sense* and *The Developing Mind* and faculty member at the Center for Culture, Brain, and Development, UCLA

"You cannot escape the power and influence of your beliefs. Pay attention to them, because they can make the difference between life and death, health and illness. *Born to Believe* brings great clarity to the emerging science of consciousness and explains how these findings about the brain mesh with certain spiritual traditions. Every thoughtful person will want to be aware of the crucial ideas discussed in this book."

—Larry Dossey, MD, author of *The Extraordinary Healing Power of Ordinary Things*

*f*P

Also by Andrew Newberg, MD

Why God Won't Go Away: Brain Science and the Biology of Belief
(with Eugene D'Aquili, MD, PhD, and Vince Rause)

The Mystical Mind: Probing the Biology of Religious Experience
(with Eugene D'Aquili, MD, PhD)

Also by Mark Robert Waldman

The Art of Staying Together

The Spirit of Writing: Classic and Contemporary Essays Celebrating the Writing Life

Dreamscaping (with Stanley Krippner, PhD)

Love Games: How to Deepen Communication, Resolve Conflict, and Discover Who Your Partner Really Is

Archetypes of the Collective Unconscious, vols. 1–4
(Shadow, Seeker, Lover, and *Healer)*

Born to Believe

God, Science, and the Origin of Ordinary and Extraordinary Beliefs

Andrew Newberg, MD
Mark Robert Waldman

previously published as *Why We Believe What We Believe: Uncovering Our Biological Need for Meaning, Spirituality, and Truth*

Free Press
New York London Toronto Sydney

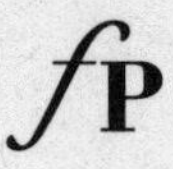

FREE PRESS
A Division of Simon & Schuster, Inc.
1230 Avenue of the Americas
New York, NY 10020

Copyright © 2006 by Andrew Newberg, MD, and Mark Robert Waldman

All rights reserved, including the right to reproduce this book
or portions thereof in any form whatsoever. For information address
Free Press Subsidiary Rights Department, 1230 Avenue of
the Americas, New York, NY 10020.

This title was previously published as *Why We Believe What We Believe: Uncovering Our Biological Need for Meaning, Spirituality, and Truth*

First Free Press trade paperback edition October 2007

FREE PRESS and colophon are trademarks
of Simon & Schuster, Inc.

For information about special discounts for bulk purchases,
please contact Simon & Schuster Special Sales at 1-800-456-6798
or business@simonandschuster.com.

Book design by Ellen R. Sasahara

Manufactured in the United States of America

1 3 5 7 9 10 8 6 4 2

The Library of Congress has cataloged the hardcover edition as follows:

Newberg, Andrew B.
Why we believe what we believe: Uncovering our biological need for meaning,
spirituality, and truth/Andrew Newberg, Mark Robert Waldman.
p. cm.
Includes bibliographical references and index.
1. Psychology, Religious. 2. Brain—Religious aspects.
I. Waldman, Mark Robert. II. Title.
BL53.N49 2006
153.4—dc22 2006048274

ISBN-13: 978-0-7432-7497-5
ISBN-10: 0-7432-7497-0
ISBN-13: 978-0-7432-7498-2 (pbk)
ISBN-10: 0-7432-7498-9 (pbk)

To Gene d'Aquili
and Jeremy Tarcher
for their vision, friendship, and relentless search for the truth.

And to our wives, Stephanie and Susan,
who stand by us night and day.

Contents

Born to Believe

Preface

GOD.

Of all the beliefs we hold—about life, the universe, everything—spiritual beliefs are the most challenging and enigmatic to study. With the growing accumulation of scientific evidence explaining human nature and cosmological evolution, one might expect that spiritual and theological perspectives would be on the decline. But this is not the case. Religion, especially in America, is flourishing, even among scientists. God simply will not go away.

Why is this so? Many theories try to explain the psychological and sociological reasons why people nurture spiritual beliefs, but the answer is found in neuroscience—indeed, in the very synapses of our brain. Simply put, we are biologically inclined to ponder the deepest nature of our being and the deepest secrets of the universe. In such states of contemplation, our brains can experience spiritual realms that feel as real as anything else we encounter in the world.

Unfortunately, our mechanisms of perception can catch only glimpses of the reality that surrounds us, and by the time these fragments of sensation reach our consciousness, we will have constructed an internal reality that is quite different from the way the world actually is. Thus, at the core of our knowledge, we find that we embrace many unconscious assumptions that never have been proven to be true.

We begin our lives without beliefs, yet our brains come equipped with a natural propensity to believe. For the first few years of existence, we unquestioningly absorb the beliefs of others—parents, teachers, friends—to help us survive in the world. We assume, quite naturally, that what we are told is true, and these basic lessons of life become our foundation for building more sophisticated beliefs and ideals.

We are born to believe in almost anything, and every child comes to know the world through a blurring of fantasies, folk wisdom, and facts. It takes decades before a child develops the capacity to question these early beliefs, which have been unconsciously imprinted into the memory circuits of the brain. In order to understand why we believe in God—or fairies, ghosts, UFOs, lucky charms, or Santa Claus—this book will take you through the stages of perceptual, cognitive, emotional, and social development that must occur to enable us to form even the simplest beliefs about life.

The brain is a stubborn organ. Once its primary set of beliefs has been established, the brain finds it difficult to integrate opposing ideas and beliefs. This has profound consequences for individuals and society, and helps to explain why some people cannot abandon destructive beliefs, be they religious, political, or psychological.

We are not born with a specific belief in God, or for that matter, any religious belief. Instead we *learn* to believe or disbelieve in God. As Richard Dawkins aptly puts it, children are not Jewish or Christian or Muslim. Rather, they are taught to believe in one set of ideas, and they are taught to disbelieve in others. With enough repetition, these beliefs and disbeliefs become neurologically embedded in memory, from which they influence future behaviors and thoughts. Thus, the more time you devote to believing in God—or making money or waging war—the more those beliefs become an integral part of your reality.

Fortunately, the neural plasticity of our brains allows us to make subtle (and sometimes dramatic) alterations to our systems of belief. Thus, when we are exposed to new ideas, we have the biological ability to alter our earlier beliefs. But we rarely abandon them fully. For Francis Collins, author of the recent book *The Language of God*, Christianity held great meaning and value in his life, but his research as a geneticist required him to immerse himself in naturalistic explanations of the universe. The result was a transformation—an evolution of sorts—of his religious beliefs. Collins maintains his faith in God, but it is not the biblical God of his childhood.

Spiritual practices also have the potential to alter beliefs dramatically. As you will discover in Chapters 7 and 8, different types of prayer and meditation can stimulate profoundly different experiences of God, and our research shows that these experiences can alter the brain in radically different ways. Our research also suggests that permanent changes can occur in the neurological circuits that monitor our conscious perception of reality. In this sense, the spiritual practitioner can actually experience a different form of reality.

But what about the person who doesn't believe in God? What happens when he or she meditates or prays? In Chapter 9, we present the first preliminary brain-scan evidence showing that, when an atheist contemplates God, a significant degree of cognitive dissonance takes place in the frontal lobes, making it difficult (but not impossible) for a disbeliever to have a spiritually uplifting experience. Take, for example, the evolutionary theorist Richard Dawkins, who has spent decades attacking religious ideologies; he has often said that he would love to have a spiritual experience, but never has. From our perspective, his disbelief makes it neurologically impossible for him to do so. The mere mention of God evokes a negative reaction in some people's brains in the same way that Judaism, Islam, or Hinduism can evoke a negative reaction in people who are deeply invested in different ideologies and beliefs. The atheist and the fundamentalist must overcome many of the same types of neurological barriers in order to appreciate the value of the other person's orientation.

Our book does not purport to prove or disprove the existence of God. Rather, we want the reader to realize how powerful any belief can be. Even more important, we want the reader to recognize that, although we are designed to have beliefs, all beliefs have limitations, and every one of them contains assumptions and inaccuracies concerning the true nature of the world. It is also important to recognize that the memories and beliefs that we have about ourselves are the most untrustworthy of all.

We also need to discern how easy it is for people to implant false beliefs in others. For example, if you listen to the media news, you might think that there is tremendous controversy raging between scientists and theologians, but a stroll through many American universities will show that, on the contrary, a deep interdisciplinary camaraderie exists. At the University of Pennsylvania, for example, we have established the Center for Spirituality and the Mind, where leaders in medicine, psychology, religious studies, pastoral care, and the neurosciences gather to share their knowledge. New research is instigated, new classes are developed, and teachers from all over the world come together to create programs that reach out to communities in need. Such interdisciplinary programs do not divide people, but bring them together.

We, the authors, do not take a negative view of religion, and the research gathered thus far shows that spiritual practices stimulate a wide range of physical and emotional benefits to the individual. In fact, very little evidence has been found showing that religious beliefs are inherently unhealthy, and even with the evidence that does exist, it is difficult to interpret it in a causally pragmatic way. Religious fundamentalism, for example, correlates highly with dogmatism, zealotry, and prejudice, yet we cannot say if involvement with fundamentalist religions promotes these socially destructive tendencies or if prejudicial individuals simply are attracted to authoritarian organizations. Also, it is important to recognize that fundamentalism is not limited to religious beliefs. For example, there are an equal number of nonreligious suicide bombers as there are religious bombers. Nor is there any evidence to support the claim that atheists are less moral than believers.

The real culprit in these cases is neither religiosity nor atheism, but the power of authoritarian individuals and groups—religious or political—to subvert the ethics of their followers. This is so essential to understand that we have devoted an entire chapter to the ease with which any person in a powerful position can manipulate good people to behave in fundamentally immoral ways.

We believe that people who engage in spiritual practices are learning how to alter neural patterns of cognition voluntarily, in ways that promote measurable degrees of happiness, compassion, and peace. Indeed, this may be religion's greatest gift to humanity: that prayer and meditation can be used to develop life-affirming goals that help people get along better with others. But religion, like politics, can be a two-edged sword, liberating some while oppressing others.

Religious beliefs have often been cited as a cause of violence, but when we look deeper into the mechanisms of the brain that direct us, we find that the forces that govern our morality and decision-making skills are far more complex than we imagined. By understanding how our brains work, we can become better believers in that we grow more aware of the inherent strengths and weaknesses of our personal truths while becoming more tolerant of those who hold different yet equally valued beliefs.

The human brain is really a *believing* machine, and every experience we have affects the depth and quality of those beliefs. The beliefs may hold only a glimmer of truth, but they always guide us toward our ideals. Without them, we cannot live, let alone change the world. They are our creed, they give us faith, and they make us who we are. Descartes said, *Cogito ergo sum*, "I think, therefore I am." But viewed through the lens of neuroscience, it might be better stated as *Credo ergo sum*, "I believe, therefore I am."

—Andrew Newberg, MD, and
Mark Robert Waldman

April 1, 2007

Part I

How the Brain Makes Our Reality

Chapter 1

The Power of Belief

Mr. Wright wasn't expected to live through the night. His body was riddled with tumors, his liver and spleen were enlarged, his lungs were filled with fluid, and he needed an oxygen mask to breathe. But when Mr. Wright heard that his doctor was conducting cancer research with a new drug called Krebiozen, which the media were touting as a potential miracle cure, he pleaded to be given treatments. Although it was against protocol, Dr. Klopfer honored Mr. Wright's request by giving him an injection of the drug, then left the hospital for the weekend, never expecting to see his patient again. But when he returned on Monday morning, he discovered that Mr. Wright's tumors had shrunk to half their original size, something that even radiation treatments could not have accomplished.

"Good God!" thought Dr. Klopfer. "Have we finally found the silver bullet—a cure for cancer?" Unfortunately, an examination of the other test patients showed no changes at all. Only Mr. Wright had improved. Was this a rare case of spontaneous remission, or was some other unidentified mechanism at work? The doctor continued to give injections to his recovering patient, and after ten days practically all signs of the disease had disappeared. Wright returned home, in perfect health.

Two months later, the Food and Drug Administration reported that the experiments with Krebiozen were proving ineffective. Mr.

Wright heard about the reports and immediately became ill. His tumors returned, and he was readmitted to the hospital. Now, Dr. Klopfer was convinced that the patient's *belief* in the drug's effectiveness had originally healed him. To test his theory, he decided to lie, telling Mr. Wright about a "new, super-refined, double-strength product" that was guaranteed to produce better results. Mr. Wright agreed to try this "new" version of what he believed had healed his tumors before, but in reality, Dr. Klopfer gave him injections of sterile water.

Once again, Mr. Wright's recovery was dramatic. His tumors disappeared, and he resumed his normal life—until the newspapers published an announcement by the American Medical Association under the headline "Nationwide Tests Show Krebiozen to Be a Worthless Drug in Treatment of Cancer."

After reading this, Mr. Wright fell ill again, returned to the hospital, and died two days later. In a report published in the *Journal of Projective Techniques*, Dr. Klopfer concluded that when the power of Wright's optimistic beliefs expired, his resistance to the disease expired as well.[1]

Each year, thousands of cases of remarkable recoveries are described, and although such "miracles" are often attributed to the power of faith and belief, the majority of scientists are skeptical of such claims. In the medical literature, spontaneous remissions—at least when cancer is involved—are extremely rare. Estimates range from one case in 60,000 to one in 100,000, although a definitive overview of the topic[2] argues that perhaps one patient in 3,000 experiences a spontaneous remission. Moreover, the majority of oncologists believe that an unidentified biological mechanism is at work rather than a true miracle[3]; and current hypotheses favor alterations in the body's cellular, immunological, hormonal, and genetic functioning over psychological mechanisms.[4] But Mr. Wright's case is unique—and one of the few to be documented during a university research project. The remissions of his cancer have been attributed to the effects of his mind on the biological functioning of his body—in other words, on the biology of belief.

Hundreds of mind-body experiments have been conducted—including placebo studies and research on the power of meditation and prayer—but few scientists have attempted to explain the underlying biology of belief. We have volumes of comprehensive statistics about the kinds of beliefs we hold, but our understanding of how and why belief "works" is still in its infancy, and most conclusions are still controversial.

Fortunately, recent discoveries about the ways the brain creates memories, thoughts, behaviors, and emotions can provide a new template with which to examine the how and why of belief. What I will propose in this book is a practical model of how the brain works that will help you understand your own beliefs and the nature of reality. It will also help you see how all beliefs emerge from the perceptual processes of the brain, and how they are shaped by personal relationships, societal influences, and educational and spiritual pursuits. This understanding can then help us to discern the difference between destructive and constructive beliefs, skills that are essential if we are to adequately address important individual, interpersonal, and global problems.

Beliefs govern nearly every aspect of our lives. They tell us how to pray and how to vote, whom to trust and whom to avoid; and they shape our personal behaviors and spiritual ethics throughout life. But once our beliefs are established, we rarely challenge their validity, even when faced with contradictory evidence. Thus, when we encounter others who appear to hold differing beliefs, we tend to dismiss or disparage them. Furthermore, we have a knee-jerk tendency to reject others who are not members of our own group. Even when their belief systems are fundamentally similar to ours, we still feel that they are significantly different. For example, Christianity, Judaism, and Islam all embrace similar notions of God,[5] yet according to one poll nearly one-third of Americans believe that each of these religious groups worships a different deity.[6] Even though a close investigation of the world's religions will show that the majority of human beings share similar ethical values, we tend to ignore the similarities and focus on the discrepancies. Ignorance is only partly to blame. A more significant reason is that our brains are instinctually prone to reject information that does not conform to our prior experience and knowledge. Simply put, old beliefs, like habits, die hard.

This book is also about our biological quest for meaning, spirituality, and truth. If we understand the neuropsychology of the brain, our beliefs will be able to grow and change as we interact with others who have different views of the world. It is my hope that as we become better believers, we will exercise greater compassion in our search for meaning and truth.

The study of human beliefs often raises unsettling issues, since most people are not aware that many of our beliefs are based on incomplete assumptions about the world. How, then, can beliefs be so powerful that they can heal us, or so destructive that they can cause us to suffer and die? This question has haunted philosophers, theologians, and politicians for a long time, and I myself have struggled to answer it for most of my medical career. For me, it all began with my own questions about the nature of reality and God.

Reality, Dreams, and Beliefs

As a teenager, I often wondered why people believed certain things. Some of my friends believed in God while others did not, but no one could give a strong enough argument to change anyone else's mind. Similar stalemates occurred when our conversations touched on issues of evolution, the origin of the universe, or more captivating topics such as basketball and girls. For the most part, our opinions (except for those about girls) never changed. In our debates, it didn't even matter what the facts were; if they didn't support our beliefs, we dismissed them. Nonetheless, I was never certain about what I should or shouldn't believe, because both sides seemed to have valid points. I knew that there was always some study, tucked away in a forgotten crevice at the library, that could support even the most outrageous claim.

By the time I finished high school, I began to think that I would never be able to know what was true or false. I even used to wonder, as teenagers are prone to do, if the world itself was real. Maybe everything was nothing more than a dream. In college, I came across the following poem paying homage to a Chinese sage born 300 years before Jesus:

Chuang Tzu dreamed he was a butterfly.
What joy, floating on the breeze
Without a thought of who he was.

When Chuang Tzu awoke, he found himself confused.
"Am I a man who dreamed I was a butterfly?
Or am I a butterfly, dreaming that I am a man?
Perhaps my whole life is but a moment in a butterfly's dream!"[7]

So I was not alone in my ruminations about reality. When I discovered that many physicists also doubt that we will ever know the true nature of the universe, I began to wonder how anyone could trust his or her beliefs. For that matter, why did people believe in anything at all? What is this impulse to believe?

Eventually, I realized that if I was to have any hope of understanding why people believe what they do, I would have to study the part of us that actually does the believing—the human mind—for no matter what we see, feel, think, or do, it must all be processed through the brain. After years of study, I have come to see that a profound chasm exists between the world "out there" and our internal consciousness, and that this fundamental disconnection prevents us from ever truly "knowing" reality. Still, we seem to have little choice but to trust our neural perceptions.

We are born to believe because we have no other alternative. Because we can never get outside ourselves, we must make assumptions—usually lots of them—to make sense of the world "out there." The spiritual beliefs we adhere to and the spiritual experiences we can have are also influenced by our neural circuitry and its limitations. God may exist, but we could experience God—or anything else, for that matter—only through the functioning of our brains.

In my previous book, *Why God Won't Go Away,* I began to address our perception of God and other spiritual beliefs by studying the brain processes that occur during meditation, prayer, and spiritual experiences. My research, conducted with my late colleague Eugene d'Aquili at the University of Pennsylvania, suggests that we are naturally calibrated to have and embrace spiritual perceptions by the neurological architecture of our minds.[8] But every individual also seems to have an abiding need to construct moral, spiritual, and scientific beliefs that explain the workings of the universe. So a belief itself is a fundamental, essential component of the human brain. As we evolved, beliefs, even superstitious ones, allowed our ancestors to make sense out of an incomprehensible, dangerous world. Their assumptions may not have been accurate, but their beliefs reduced their fears and imparted values that would facilitate group cohesiveness.

Prejudice, Skepticism, and Doubt

The propensity to believe that other people's values are misguided has fostered centuries of animosity throughout the world. When the early Christian missionaries first observed shamanic rituals practiced by indigenous tribes outside Europe, they usually thought of these rites as the devil's work. They believed that punishment and conversion were essential for the salvation of the natives' souls. The French Franciscan priest André Thevet, when visiting Brazil in 1557, noted in his diary:

> I cannot cease to wonder how it is that in a land of law and police, one allows to proliferate like filth a bunch of old witches who put herbs on their arms [and] hang written words around their necks . . . to cure fevers and other things, which are only true idolatry, and worthy of great punishment.[9]

How would such priests react today if they were to wander down the aisles of an American health-food store filled with exotic tinctures and herbal preparations? The sheer numbers of Protestants alive would no doubt make them long for another Inquisition.

Neurologically, such prejudice seems rooted in human nature, for the human brain has a propensity to reject any belief that is not in accord with one's own view. However, each person also has the biological power to interrupt detrimental, derogatory beliefs and generate new ideas. These new ideas, in turn, can alter the neural circuitry that governs how we behave and what we believe. Our beliefs may be static, but they aren't necessarily static. They can change; we can change them. Nowhere is this more apparent than in the workings of a child's mind, which is constantly struggling to develop and maintain a stable worldview. Furthermore, children's and adults' belief systems are continually being altered by other people's beliefs.

The adult human brain is childlike in another way: we automatically assume that what other people tell us is true, particularly if the idea appeals to our deep-seated fantasies and desires. Advertisers often take advantage of this neural tendency, and even though consumer advocates and some laws have helped to level the playing field, the general rule "Buyer beware" still prevails. Magazine covers and full-page ads promise instant beauty, fabulous sex, and intimate communication in five easy steps, and we believe them, often ignoring obvious deceit. One ad I recently saw—in a popular science magazine, no less—promised the reader a complete aerobic workout "in exactly four minutes": a medical impossibility, at least from the standpoint of cardiovascular health. So how does the advertiser get away with this? Through a definitional loophole. Technically, "aerobic" simply means that a certain activity provides oxygen to the system, so any movement—even rolling around in bed—would bring oxygen to any muscle that moves. There is little health benefit to this, but the ad tricks you into thinking that you get the same benefits as if you had exercised vigorously for twenty minutes or longer. Furthermore, the advertisers like this one are preying on many people's propensity to want quick, efficient solutions that require little effort.

Food manufacturers present their products in similar ways. For example, many labels state that the ingredients in a product are "all natural." As far as the Food and Drug Administration is concerned, this simply means that the product contains no metal, plastic, or other synthetic material. "Natural" does not mean "healthy" or "organic," but as advertisers know, such pseudoscientific jargon can dramatically increase a product's sales.

We are born with a natural tendency to trust what others say, and we certainly can't take the time to question every piece of information we receive. Think how long it would take to verify even half the claims that are made in just a single magazine. So what are we to do?

One thing we can do is train ourselves to become more vigilant and cautious. Adopt a skeptical, open-minded attitude. I'm not recommending that you become a pessimist—unfortunately, many people incorrectly equate skepticism with pessimism, doubt, and disbelief. Philosophical skepticism dates back to the time of Plato, who established the first school of "academics," teaching that the world could not be known objectively or precisely. The academics also believed that the true nature of God could never be fully known. Thus a skeptic is simply a person who chooses to examine carefully whether his or her beliefs are actually true. A skeptic keeps an open mind—a willingness to consider both sides of an argument. In reality, we need a healthy dose of skepticism, open-mindedness, and trust, especially when it comes to those beliefs of our own that affect another person's life. This is particularly important with regard to assumptions we make in medicine and science, and it is also important when we are addressing moral, political, and religious issues. Trust and open-mindedness without some skepticism can get us into trouble, but skepticism without trust can undermine our ability to believe what we need to in order to survive. Each has its benefits and risks. For example, as Carl Sagan once pointed out, the business of skepticism can threaten the status quo:

> Skepticism challenges established institutions. If we teach everybody, including, say, high school students, habits of skeptical thought, they will probably not restrict their skepticism to UFOs, aspirin commercials, and 35,000-year-old channelees. Maybe they'll start asking awkward questions about economic, or social, or political, or religious institutions. Perhaps they'll challenge the opinions of those in power. Then where would we be?[10]

Although Sagan was being ironic, skepticism can be taken too far. It can cause us to reject out of hand new ideas that, on the surface, seem improbable or weird. It can also lead to cynicism, a state in which one constantly doubts the sincerity and validity of another person's point of view. And this, as every psychiatrist and cardiologist knows, can lead to anger, bitterness, contempt, hostility, and depression. In the long run, the hormonal and neurological changes caused by these emotional states can seriously compromise physical health.

How, then, do we know whom, or what, to trust? And how do we keep an open mind, particularly when we encounter claims that contradict our personal experience and faith? Science explains that the universe is billions of years old, and that human beings and chimpanzees have evolved from a common ancestor, but reactions to this information still range from skepticism and cynicism to open-mindedness and acceptance. It is not easy to challenge assumptions that have prevailed for hundreds, even thousands, of years.

Science, Medicine, and Faith

In medical research, I feel it is wise to be skeptical about new treatments because we are dealing with people's health and lives. I need to see a good amount of persuasive evidence and data before I'm comfortable trying a new procedure. However, my skepticism can ultimately lead to my becoming open-minded and trying a new treatment, which can lead to better health for my patients. If I were to apply this clinical skepticism to everything, I'd be living in a constant state of doubt, which is a very inefficient way to live on a day-to-day basis. Marriage is a perfect example: at some point in every intimate relationship we have to abandon our doubts and believe that our partner will continue to be trustworthy in the future. In other words, we have to have faith in ourselves, and in other people with whom we interact regularly, especially those we love.

Mr. Wright had to have faith in his doctor, and Dr. Klopfer had to have faith in the power of his patient's belief. Such faith transcends reason, rationality, and skepticism, and has the power to heal, but there is nothing magical about it. In fact, you can evoke placebo effects in mice and other animals.[11] The truth and measurability of the placebo effect allow us to begin to trace the neurophysiology of belief. Essential elements in the construction of any type of belief[12] include the mechanisms of perception, appraisal, attention, emotion, motivation, conditioning, expectancy—and, in the case of human beings, verbal suggestion. Fear, anxiety, and doubt also contribute to the placebo effect, but in a negative way, creating disbelief that can interfere with the healing processes of the body. In Mr. Wright's case, we can see both types of belief operating in profoundly powerful ways. Without any evidence or proof, he became convinced, beyond reason, that he would survive, and this strongly held expectation seemed to play a significant role in reversing the progress of his disease. Most likely, his brain sent out chemical signals that stimulated his immune system, in ways that we are just beginning to understand. Then, when he read reports that the medication didn't work, his emotional despair, coupled with the negative belief that he was bound to die, turned off his immune response and simultaneously released a flood of stress-related hormones, some of whose effects we do understand.

Other factors probably played essential roles in the roller-coaster course of his cancer. For example, studies have found that injections of harmless substances—even water—can trigger the suppression of tumors in rats (this is known as a learned immunosuppression response), but there is also evidence that these conditioned rats have a weaker ability to resist tumors that occur at a later date.[13] This may indicate that our positive beliefs might help to postpone the inevitable decline of health. To me, this is an amazing finding, for if future research supports this hypothesis with humans, it means that, one day, we might learn how to control our minds to extend both the quality and the quantity of our lives.

It is also my conviction that the more we understand the biological underpinnings of belief, the easier it will be for a person to come to a middle ground between blind trust and the blanket rejection of anything that seems foreign or strange. However, we will still be faced with the problem that we cannot get outside our brains to know what reality is, and so we must live with the paradox that there may be no clear delineation between fantasy and truth.

Do All Living Organisms Hold Beliefs?

What about other creatures? Do they, like humans, have beliefs? It depends on the kind of brain they have. Recent studies have revealed that primates and other animals do form rudimentary beliefs about their world. For example, many wild creatures can be trained to trust a human being, and this demonstrates their ability to form new assumptions about their environment. Dogs, for example, will sit expectantly by the front door for hours, waiting for their owners to return. In fact, most canines are inveterate, optimistic believers in the goodwill of their masters. My dogs can even anticipate the time of day when I am supposed to return home, and will begin to react by barking the moment I call to let my family know that I'm on my way. At any other time of day, they do not bark when the phone rings. Biologically speaking, this illuminates the processes of belief that are involved in maintaining faith about a projected future event.

Even the behavior of single-celled creatures can be conditioned and changed. When an amoeba is gently shocked with an electric probe, for instance, it becomes more hesitant when exploring its surroundings: it no longer assumes that the world "out there" is safe. In a manner of speaking, you might say that the otherwise trusting amoeba becomes a skeptic. If the shocks continued, it would probably turn into a hermit, retreating from its environment until it died. If you think this scenario sounds too improbable, consider the *Dictyostelium discoideum,* which biologists affectionately call the social amoeba. This little creature exhibits what appear to be moral behaviors, for it engages in cooperative activities that involve both cheating and altruism.[14] If enough evidence is gathered to support the view that cells and genes can independently and cooperatively make decisions that affect their own future survival, then the answer is yes—every living organism has beliefs.

What about rock? It has no nervous system or cells, but is there even a remote possibility that the smallest subatomic particles of the universe could have some form of self-volition, consciousness, or belief, which would then suggest that the universe itself is a form of life? Most quantum physicists would say no.[15] However, a few respected theoreticians and physicists believe that it is impossible to separate consciousness from the physical world, and that a profound interconnectedness exists between all aspects of the organic and inorganic world. For example, the Gaia hypothesis proposes that every aspect of the environment on Earth cooperates in a self-regulating way to maintain an internal and external balance.[16] There is even a mathematical theory explaining how two species of daisies can regulate the global temperature of our planet.[17]

Interestingly, the notion that inanimate objects have a kind of consciousness is reflected in the myths and spirituality of many indigenous cultures. Believing that everything—rocks, trees, and the heavens above—has consciousness, they feel more connected to the mysterious world out there. Charles Alexander Eastman, who was a Santee Sioux and a physician, expressed this when he wrote, in 1911, about the continuing annihilation of the Native Americans' way of life:

> We believed that the spirit pervades all creation and that every creature possesses a soul in some degree, though not necessarily a soul conscious of itself. The tree, the waterfall, the grizzly bear, each is an embodied Force, and as such an object of reverence. The Indian . . . had not yet charted the vast field of nature or expressed her wonders in terms of science. With his limited knowledge of cause and effect, he saw miracles on every hand—the miracle of life in seed and egg, the miracle of death in a lightning flash and in the swelling deep![18]

Science cannot yet verify the existence of consciousness beyond the brain, but we do have evidence that such beliefs can generate a sense of peace and equanimity within the brain: the more connected we feel with the world, the more empathy we express toward others. This sense of connection may even be neurologically essential for the development of moral ideals.

To summarize, our beliefs serve myriad purposes that help us to flourish and survive:

- They help us to organize the world in meaningful ways.
- They give us our sense of ourselves.
- They help us take action in specific ways.
- They allow us to accomplish our goals.
- They help to regulate the emotional centers of the brain.
- They allow us to socialize with others.
- They guide us in our moral and educational pursuits.
- They heal our bodies and minds.

Beliefs can also be used to suppress others, to justify immoral acts, or to propel us toward sadistic acts. But if we so choose, they can also connect us with transcendent dimensions of experience, be it seen through religion, science, or the innate curiosity of a child's imagination. Most important, they can give us inspiration and hope, essential tools for confronting those moments of confusion and doubt that are so often part of life.

Chapter 2

A Mountain of Misperceptions: Searching for Beliefs in a Haystack of Neurons

IN THE HARRY POTTER NOVELS, HOGWARTS IS A SCHOOL OF magic for girls and boys in which they have classes in Potions, Divination, and Defense against the Dark Arts—courses that are not entirely different from those I had to take in medical school. As medical students, we'd write prescriptions for pharmaceutical potions that act on the brain in magical ways (we still don't know, for example, how antidepressants actually work), and we'd attempt to make sense of the blotches and blurs from brain scans, which at first seemed more like reading tea leaves in a divination class. And our defense against the dark arts would be the practices we developed to keep awake after thirty-six hours of being on call.

Like Hogwarts, the brain is filled with hallways, labyrinths, and hidden rooms that shift their direction when you least expect them to. The brain is continually rearranging the cognitive information it stores, divining new meanings and beliefs with every experience it perceives. One of the characters in the Potter books was uncomfortable with divination because it was unpredictable and vague, and we neuroscientists can feel the same frustration when examining the processes of human perception. You can't pinpoint a memory, or surgically remove a belief—these functions are like ghosts in the machinery of the brain. Different beliefs emerge in different parts of the brain at different times and under different circumstances, and they are also influenced by factors that occur outside the brain. Ideas, thoughts, and feelings are difficult to research because they are end products of complex neural processes that include perception, emotion, memory, and behavioral motivation. The moment you begin to define a belief, a bunch of other concepts—such as awareness, cognition, and consciousness—crop up that are equally difficult to explain.

Looking for a belief, even if you're using the most sophisticated brain-imaging technology, is like looking for a needle in a haystack. To use a religious metaphor, it's like looking for God in the universe. God is everywhere and nowhere, depending on whom you ask, and the same holds true for beliefs: they seem to be everywhere and nowhere within the brain, again depending on whom you ask. Philosophers like Daniel Dennett and Lynne Rudder Baker even suggest that the notion of beliefs is not scientifically valid. However, Dennett and Baker argue that treating people as if they had beliefs is a useful strategy for understanding human behavior.[1]

For those who study the nature of human consciousness, beliefs seem like a sorcerer's apprentice who is constantly playing tricks with our mind. And yet beliefs are our most important human commodity. With them we can build civilizations, make revolutions, create music and art, and determine our relationship to the cosmos. Beliefs make us fall in love, and they drive us into hate; that is why it is so critical to understand how they work. We all have beliefs, we all need them, and they will determine humanity's fate.

Religious and spiritual beliefs have had a particularly profound influence over human history, and yet we barely grasp how they work at the biological, behavioral, or psychological level. As a neuroscientist, I have come to realize that the study of beliefs may be the single most important quest, both scientifically and spiritually. Furthermore, I think we must begin this exploration by examining the very part of us that does the believing—the human brain.

With over 100 billion neurons to study—each having up to 10,000 dendrites to connect with other neuronal structures—we scientists might never be able to figure out how that mass of gray matter works completely. But this possibility won't stop us from trying. We'll build theories and hypotheses about the inner workings of the brain: educated guesses based on the limited information we have gathered. In a similar manner, our brains make educated guesses about the true nature of the external world by drawing maps and making elaborate assumptions and predictions about future outcomes. In other words, the brain is busy constructing inner beliefs about the outer workings of the world. Sometimes we get it right, and sometimes we don't. Fortunately, the human brain comes equipped with a very special feature: it can alter its system of beliefs far more rapidly than that of any other organism on the planet. Thus beliefs act as an invisible but intelligent inner pilot guiding the complex activities of our lives.

A belief is like a map, a neural representation of an experience that seems meaningful, real, or true. It begins with the first hints of information coming in through our senses, and it culminates in the nebulous territory called consciousness. In the process, billions of synaptic processes transform neural data into categories, concepts, emotions, memories, language, thoughts, and knee-jerk reactions to a broad assortment of stimuli ranging from the innocuous and pleasurable (like blue skies and apple pie) to the noxious and disagreeable (like spiders or politicians). But the map is not the territory. It's an abstraction, a symbol of something that we assume exists, like a lamppost or a feeling of satisfaction. We may not have any direct evidence or proof of its existence, but we do have this great internal map, and for the most part, it appears to work quite well.

Our brain also makes our internal map seem real. Even schizophrenics believe in the reality of the voices they hear, because the brain has few options but to rely on the maps it makes. We do the same thing when we're driving around in a car. We take our *Thomas Guide* or our GPS navigation system for granted, and we believe that if we follow those abstract squiggles and numbers, we will end up at our friend's house rather than at the city dump. But the lines, squiggles, and numbers on a map are not the same as the roads we drive to reach the house. They are two-dimensional representations of a three-dimensional world.

Instead of paper, the brain uses memory; and instead of ink, the brain turns on circuits. And the three-dimensional world we think we perceive? It's really our imagination at work, for we never actually "see" the world directly. The brain takes the raw information—consisting of lines, shapes, and contours—that activates cells in our eyes and creates a representation of the room around us with chairs, tables, and doors, so that we can get up at some point and walk out of the room without crashing into anything. The vivid three-dimensional world that we are conscious of is created by neurochemical and neuroelectric impulses that take the world "out there" and make a picture inside the human brain.

> To a very large extent men and women are a product of how they define themselves. As a result of a combination of innate ideas and the intimate influences of the culture and environment we grow up in, we come to have beliefs about the nature of being human. These beliefs penetrate to a very deep level of our psychosomatic systems, our minds and brains, our nervous systems, our endocrine systems, and even our blood and sinews. We act, speak, and think according to these deeply held beliefs and belief systems.
>
> **—Jeremy W. Hayward, author and physicist**

Unfortunately, imagination, memory, and consciousness are not very stable mechanisms. Even the "wiring" of our neural circuits continues to form and change as we acquire new experiences and beliefs. That is why the things we first observe are a bit different each time we call them into consciousness. For example, although we're not aware of it, we have altered and embellished our childhood memories so many times that some of the events we recall may never have happened at all. And since there is a huge gap between the world out there and our inner worldview, the brain stays busy revising its cognitive maps, selecting some perceptions, ignoring others, and filling in the blanks with conjecture. However, the brain can help us detect perceptual and cognitive discrepancies—for example, we're very good at detecting lies and deception. Beyond that, the brain tends to trust its intuitions about the world. These intuitions are the neural equivalent of beliefs.

Defining Beliefs

In the neurosciences, we strive to define our terms as accurately as possible so that, at the very least, other scientists will understand us. Unfortunately, subjective experiences such as feelings, values, and meaningfulness are difficult to define because they mean different things to different people. Unless we clarify our terms, we cannot come to a consensus on which to base our observations and experiments.

Think about it for a moment. How would you define "belief"? Usually, when I am asked to delineate a difficult term, I turn to my two favorite sources of inspiration: my six-year-old daughter, and a dictionary. First, my daughter. When I asked her for a definitive statement about beliefs, she said, "That's easy! A belief is something I believe in." Adults, too, have been known to use such circular logic, particularly when the answer seems obvious. After all, we usually take our notions of belief for granted.

My other source, the *Oxford English Dictionary*, defines "belief" in the following ways:

1. A feeling that something exists or is true, especially one without proof.
2. A firmly held opinion.
3. Trust or confidence in.
4. Religious faith.

This definition, like most definitions, distinguishes between things that can and cannot be proved. Many people have used this distinction to argue that religious beliefs are flawed. These critics often fail to recognize, however, that what constitutes a proof about anything is also a form of belief, and is based on rules that are themselves filled with unproved assumptions. It's circular thinking on a grander scale; and as I will soon explain, proofs are never exempt from errors. A proof, according to Webster's dictionary, is any sequence of steps, statements, or demonstrations leading to a valid conclusion. However, different fields of thought (philosophy, science, law, etc.) apply different standards for establishing facts, and this is where conflicts emerge between religious and scientific perspectives. What satisfies one person in a proof of God's existence may not satisfy another person who is applying a different set of rules. A theologian may have faith that a mystical vision is a gift from heaven, while a neuroscientist may swear that it was merely an electrochemical surge in the temporal lobe. Ultimately, the system of beliefs that any person will come to embrace is the one that brings the most comfort and makes the most sense.

Synonyms for Belief

Opinion, conviction, confidence, faith, trust, assumption, expectation, certainty, persuasion, assurance, acceptance, doctrine, dogma, tenet, principle, creed, supposition, attitude, allegation, knowledge, interpretation, representation, judgment, argument, advice, estimation, passion, sincerity, hope, theory, premise, possibility, probability, conjecture, hypothesis, worldview, guess.

So how might we define belief in a way that allows us to study it scientifically? Biological and neuropsychologically, a belief can be defined as any perception, cognition, or emotion that the brain assumes, consciously or unconsciously, to be true. Throughout this book, I will use the term "perception" to refer to the information we receive about ourselves and the world through our senses. "Cognition," however, represents a different level of processing within the brain, and includes all the abstract conceptual processes that our brain uses to organize and make sense of our perceptions. Memories and consciousness are part of cognition, but as I will explain later, dozens of other cognitive activities are also essential for building beliefs. Emotions play a distinctly different role in neural processing, and help to establish the intensity and value of every perceptual and cognitive experience we have. Finally, every person's belief system is influenced by the input he or she receives from other members of the community, for if we do not experience adequate social consensus, many of our most cherished beliefs would never emerge into consciousness.

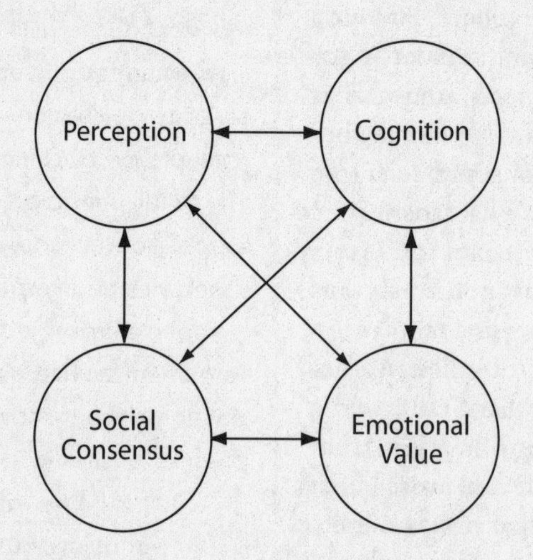

Together, these four interacting spheres of influence—perception, cognition, emotion, and social consensus—allow us to identify, explore, evaluate, and compare a wide variety of beliefs, from our most mundane evaluations of the world to the most extraordinary visions that illuminate our purpose in life. These influences affect the strength, power, and relative truth of a specific belief. Each circle of influence has a "volume control," and the greater the overall volume, the more real and truthful that belief becomes. For example, if a stranger walks up to you and mumbles something that you can't understand, it will have little emotional value. If the stranger loudly announces, "You're a millionaire," you'll certainly have an emotional reaction, but your cognitive skills (particularly your recollection of your $5,000 credit card bill and your recent bank statement showing a balance of $12) will probably persuade you that the stranger is lying. If the stranger hands you a certified check, you will probably have a stronger emotional reaction, but it still won't make much sense. Instead, your cognitive processes of disbelief will kick in; you will wonder who might be pulling your leg. But if the degree of social consensus increases—your bank tells you the check is valid—then you will begin to believe you are rich. And if the stranger turns out to be an attorney executing the will of your long-lost billionaire uncle, then all the pieces—perception, cognition, consensus, and emotional gratitude—will come together. You'll finally believe you're a millionaire, and you'll be thrilled and happy, until the IRS shows up at your door.

This model suggests that a person who has not had a strong religious or spiritual experience might have trouble believing in God. But if, in childhood and adulthood, you were surrounded by people who held deep religious beliefs, then the sphere of social influence could compensate for your own lack of perceptual experience. If you then immersed yourself in spiritual literature, the strength of your cognitive beliefs would grow, and this growth would emotionally affect your brain. Still, you'd have to find personal value in such thoughts before a sense of spiritual reality took hold. If you felt no such value, you would be far less likely to believe.

Why Should We Believe Anything at All?

Over the centuries, many pundits and sages have told us what and how to believe, especially regarding things that we cannot directly perceive with our senses. Many use logic and persuasion to convince us of their truths, but if we can't see something, and if there is no substantial evidence of its existence, then why should we take someone else's word for it? In fact, why should we believe anything at all? And yet we do believe many things. We take our parents' word for the truth; we trust the news; we accept the opinions of our friends; and we believe in all sorts of things—like love—that seem to have no substance in the world. We can't see love, yet nearly every-

> Of all the beliefs you have, which one would be the most disturbing for you to give up, if you found out that it wasn't true?

one believes in its power. Associated ideas such as romance and passion are also widely held. Do these ideas exist anywhere outside our conscious imagination?

As neurological evidence accrues, the answer is leaning toward "no." Biologists begin by searching for evidence of these emotions among other living species, but first they have to define what they mean, in a way that can be experimentally tested. If you define love as a form of nurturance and attachment, then yes, you'll find such behavior in many species. But if you define it as the falling-head-over-heels experience that every adolescent yearns for, then no, there is little evidence that other animals feel such passion. In fact, the sex life of most living organisms is dull and brief. Monogamy—an ideal behavior in many human cultures—appears in nature among only a few species, such as the jackdaw, the dik-dik, and a few kinds of termites. "True monogamy is rare," says Dr. Olivia Judson, a research fellow at Imperial College in London. "So rare that it is one of the most deviant behaviors in biology."[2]

But you won't be able to convince your teenage daughter that love—especially true love—is a figment of her imagination, especially when it is supported by the shared fantasies of millions of other adolescents (i.e., social consensus). Love seems real because the emotions triggered by a combination of hormones and romantic ideals are very powerful and often impart a strong impression of reality.

When biologists study human sexual and mating behaviors, they generally concede that love is a belief existing primarily inside one's mind.[3] I'm not saying that love doesn't exist. I'm saying only that love is a conceptual and emotional belief, far removed from the atoms and molecules that make up the physical dimensions of life. Like many of our other beliefs and ideals (democracy, freedom of speech, etc.), love may not physically exist in the world outside the mind. Still, it has the power to alter the course of our lives, and even to change the course of history.

Why, in the complex biology of the human body and the human brain, do we build abstract systems of unproved beliefs? The simple answer is that we have no choice but to believe. From the moment we are born, we depend on others to teach us about the world. As children, we are given a specific language, a particular religion, and a taste of science, and we unconsciously assume that we are learning facts about the world. We are not. We are simply being told what to believe. For the most part, this system is practical because a young child cannot perceive many of the dangers hidden in life's activities. Without guidance, children would walk into traffic, eat out of the dog's bowl, or poke their baby sister in the eye. And so our parents use everything at their disposal—threats, wisdom, punishments, rewards—to convince us of certain things. They reinforce these teachings by telling us that bad things will happen if we don't believe: our teeth will fall out (if we don't brush), God will punish us (if we don't pray), the moral fiber of the country will go to hell (if we don't vote for the candidate they support). And they'll enlist others—friends, dentists, politicians, priests—to reiterate these beliefs. Sometimes they'll succeed, and sometimes they won't. I, for example, cannot convince my six-year-old that she needs to brush her teeth. Of course, I probably don't help matters by leaving a $5 bill under her pillow whenever she loses a tooth. For all I know, she's hoping to make a fortune by losing all of her teeth.

Ultimately, each person is free to choose which beliefs to accept and which to reject. As a result, there are about 6 billion belief systems in the world. No two are identical, and yet many are much the same. We may modify some of our beliefs as we go through life; but the older we get, the less they will change, in part because of the architecture of the aging brain. Still, no matter how old we are, we need our beliefs to get us through the day.

Measuring the Power of Belief

The English word "belief" first appeared in the twelfth century, when it was adapted from the German *gilouben*, "to hold dear" or "to love." At first it was used in conjunction with religious doctrines held to be true, referring to one's trust and faith in God.[4] Faith, rather than fact, is the key word here, since the existence of God cannot be tested or subjected to the rigorous proofs developed by science. Still, the scientific method of devising falsifiable hypotheses, and then gathering data to support or reject them, undermined many theologies that were firmly established in fourteenth-century Europe. The idea of God's universe—with Earth at its center—began to collapse because some of this concept was inconsistent with accumulating evidence. But what happens if we can't definitively prove or disprove a certain idea or belief? We simply return to the fundamental mechanisms used by our brains: perception, cognition, social consensus, and, perhaps most important, an intuitive feeling of what seems right. If an experience or idea doesn't make sense, and if it doesn't feel good, then we probably won't build a very strong belief system around it.

Recently, science has made initial strides into the murky landscape of beliefs. By using the results of research into brain injury together with those from experiments with animals and experiments in the social sciences, we have begun to chart the neural processes that distinguish fantasies, emotions, and facts. With the aid of brain imaging technology, we can actually watch neural activity when a nun prays to God, or when a person encounters information that is discrepant with his or her belief. These findings are useful, but they also require a lot of interpretation, and interpretation is, in a sense, another word for belief. A picture may be worth a thousand words, but a single brain scan can generate a dozen hypotheses and an equal number of doubts.

We may not be able to take a picture of a specific belief, but we can record the traces that it leaves behind: we can see the emotional response that a belief triggers. In this way we can begin to evaluate the biological effects that a particular idea has had on a person. For example, a photograph of a tragic scene will evoke different neural reactions in different people, and these reactions can be correlated with specific beliefs the individuals hold about violence, suffering, or death.

Nonetheless, we are in only the earliest stages of learning how to map the functions of the brain. In fact, most of the imaging studies we will refer to in this book measure neural functioning on a relatively crude scale. When we look at a quarter-inch section of the brain with a scanning device, the area we see may contain thousands of neurons, and it is quite possible that only a small percentage are active during a specific thought. It's very difficult to measure, particularly since that quarter-inch matrix is connected to many other quarter-inch segments of the brain, any of which may be related to the functioning of the tiny area we are observing.

At present, neuroscience is more of an art than a science, particularly in the way it evaluates complex mental processes. It is filled with assumptions, conjectures, postulates, and rationalizations. That's why scientists demand multiple peer-reviewed studies before accepting something as factual. And even 1,000 studies will leave a degree of doubt, particularly since the same evidence can fuel opposing interpretations and conclusions. Yet we scientists do the best we can with the information we discover, even though we may never grasp the full truth about anything, because, as I will explain in Chapter 3, our primary mechanisms of perception, by their very nature, distort the reality that exists "out there," beyond the brain.

Conscious and Unconscious Beliefs

Other conundrums arise when we try to analyze the nature of human consciousness. Our values and ethics are clearly beliefs, but does a religious belief stimulate different parts of the brain from, say, a political or romantic belief? Common sense says yes, but vastly different beliefs can share similar neural circuits. In my own studies, for example, I've seen that a Buddhist meditator, who does not embrace a western notion of God, still evokes some of the same neural pathways when focusing on a sacred Tibetan object as a Franciscan nun who is focusing on a passage from the Bible. The beliefs are worlds apart, but the inner experience is often the same. In fact, contemplative and transcendent states are, in some ways, very similar to a person's pleasurable experience of music, sex, or good food. This suggests that many activities have common circuits that we might consider spiritual, religious, or just plain fun. Nonetheless, important neural differences also exist between different experiences, traditions, or practices.

These differences raise an intriguing question: is it possible that some of our beliefs function within a neural realm that is separate, even divorced, from other processes of the brain? Recent evidence supports this hypothesis, which requires us to distinguish between beliefs that are related to sensory and perceptual processes and those that are constructed out of the more abstract conceptual processes taking place in the cognitive centers of the brain. In fact, our perceptions of reality are completely transformed into abstract packets of information that are as far removed from the perceptual, behavioral, and orientation processes of the brain as the brain is from the world.[5] This new conceptual reality will be further processed, until a very small part of it bursts into consciousness. This is the reality that we become aware of and use to interact with the world around us.

Memory, Consciousness, and Emotional Realities

As the brain builds its conscious map of reality, a wide range of emotional responses will be assigned to everything we observe and think about. Even when we watch a horror film, and we know it's make-believe, parts of our brain react as if it were real, and for a moment we react with fear. In neurological terms, as we watched the movie, the limbic system—the primary emotional controller in the brain—became very active and fired off a response to other parts of the brain. This reaction is like an emotional fingerprint, and can be measured with brain-scan technology and compared with other people's responses. These data allow us to map some of the terrain of emotional beliefs. If the emotional response is strong enough, it will leave another neural imprint in the form of a memory, and any similar perception or idea that is later experienced will reignite the activity in that circuit, bringing with it aspects of the original emotion and other related memories. Certain researchers argue that memories should be considered a form of neural belief, since there is no way the brain can prove what did or did not happen in the past. "And because memory is a fundamentally constructive process that is sometimes prone to error and distortion, it makes sense that such beliefs are occasionally misguided."[6]

Most of the brain's activity involves our perceptions of the world and the internal state of our body, and these processes are primarily unconscious. Any conscious awareness of the maps we are making occurs sometime after the event takes place—between one-tenth and one-half a second later, to be exact. That's a long time to wait for the brain to tell you what's going on, particularly if a lion has just walked into your cave. This lag time is additional evidence that consciousness is many steps removed from the brain's perception of reality. Fortunately, the brain is designed to react to danger before conscious control kicks in.

In the following drawing, try to imagine that reality is everything inside the box, and that the world, as we perceive it, includes everything that exists outside the brain.

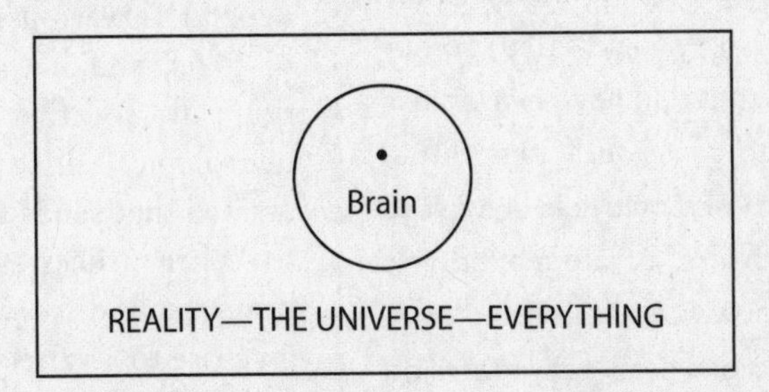

A more accurate picture of reality would be a three-dimensional container that is a quadrillion times larger; but to keep the publishing costs low, I've used this small two-dimensional box. The circle is your brain, and the dot is your consciousness, your moment-to-moment awareness of what is happening to you in the world. The brain can capture only a minuscule amount of the universe; and your consciousness can glimpse—and hold for about thirty seconds—only a very small fraction of what the brain perceives.*

Within that dot of consciousness is a microscopic molecule that symbolizes our capacity for language-based communication. Daniel Dennett, a professor of philosophy and director of the Center for Cognitive Studies at Tufts University, views all forms of communication as a series of expressed beliefs, intentional propositions that one person wishes to convey to another for various reasons.[7] Thus most of the beliefs we are consciously aware of are defined by and limited to the rules that govern language. But consciousness is based

* Some researchers have argued that consciousness occurs throughout the brain on a much broader scale; others have argued that consciousness may actually be the sum total of all the brain's activities. If such theories turned out to be true, then we'd have to fill the circle in the drawing with millions of additional dots.

The Limitations of Consciousness

The brain is limited with regard to how much information it can perceive and store. In 1956, George A. Miller presented the idea that short-term memory (which is the information we need to have available in order to consciously attend to a specific task) could hold only five to nine "chunks" of information. This idea is now well substantiated; however, no one has clearly identified what a "chunk" consists of. It can refer to digits, words, simple images, or larger conceptions like democracy or love. Here's how chunking works. Look at the following sequence, close your eyes, and try to recall the numbers: 1 – 2 – 1 – 5 – 5 – 5 – 5 – 4 – 6 – 5 – 7. You probably did not succeed. If, however, I regroup these eleven numbers into four smaller "chunks" of information, you'll have no trouble remembering 1-215-555-4657 as a telephone number. This is one of many ways that our brain compensates for its limited capacity to grasp reality.

on a different model of logic from other parts of the brain (we'll be discussing some of these cognitive models of logic in Chapter 3).

All beliefs—perceptual, cognitive, and conscious—depend on various systems of logic, and if the pieces don't fit together well, a neurological dissonance is created that sends an alarm to other processes in the brain. Such dissonance can give rise to a variety of disbeliefs. With so many gaps between reality and perception, between perception and cognition, and between conscious and unconscious thoughts, it is amazing that we believe anything at all. And yet even with these limitations, the brain does provide us with a clear, coherent sense of reality.

Our consciousness also does something remarkable: it takes the few perceptions that we hold, ignores the discrepancies, and turns them into sophisticated visions and inventions, something no other living organism can do. The visions eventually become part of the reality of the brain, and the inventions become part of the world.

Emotions Make Thoughts and Perceptions Real

Another element essential to beliefs is the meaning—or value—bestowed on the individual belief. "Value" refers to the importance or worth assigned to a particular perception or idea, and the activity of assignment can be traced to the emotional circuits of the brain in the limbic system. In general, as neural stimulation increases in these areas—which include the amygdala, the thalamus, and the hippocampus—perceived value increases. Without this activity, emotional memories—including experiences such as anger, sadness, happiness, disgust, and surprise—could not be formed, and life would have little meaning.

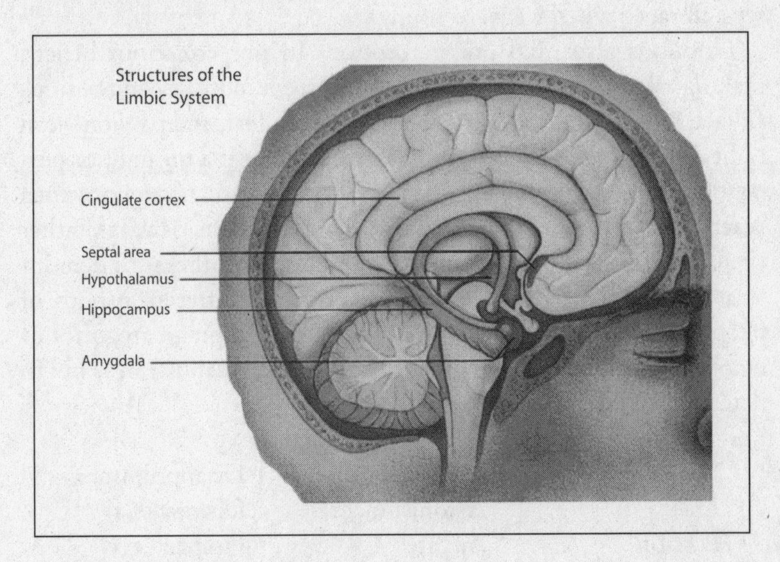

Epileptics who have had their amygdala removed to reduce unbearable seizures may have substantial impairment of their ability to respond with negative emotions. They will not be able to assign value to various events, especially events that typically evoke fear. By contrast, a person with an overactive amygdala will often live in a constant state of fear or anxiety. Such individuals tend to believe that everything is a potential threat, although if they train themselves—through meditation or psychotherapy—to focus consciously on the belief that everything is safe, the activity in the amygdala can be inhibited, thus extinguishing, at least temporarily, the feelings of anxiety and fear.

The hippocampus, another part of the limbic system, plays a major role in regulating our emotions by helping to balance the fear or anxiety response in the amygdala. The hippocampus also utilizes emotions to help establish long-term memory. Thus very emotional events tend to be written into memory more strongly than nonemotional events. As we shall see later, it is also relatively easy to implant false memories, especially in children. Furthermore, there is mounting evidence that the brain maintains false memories for extended periods of time. All this suggests that we should be very cautious about assuming the truthfulness of our beliefs, especially those that we embraced when we were young.

Emotions also bind our perceptions to our conscious beliefs, making whatever we are thinking about seem more real at the time. In fact, strong emotions—particularly anger, fear, and passion—can radically change our perceptions of reality. But if a thought or perception does not stimulate an emotional response, it may not even reach consciousness. By looking at belief in terms of value rather than truth, a scientist can formulate and test hypotheses to demonstrate which beliefs hold emotional value for different groups of people. Researchers have devised many overlapping categories of emotional beliefs and feelings, as the following list illustrates,[8] but to date few have been neurologically investigated.

Love	Amazement	Disappointment
Lust	Astonishment	Dismay
Passion	Anger	Despair
Longing	Irritation	Hopelessness
Attraction	Optimism	Displeasure
Affection	Rapture	Shame
Adoration	Relief	Guilt
Fondness	Surprise	Gloom

Tenderness
Compassion
Sentimentality
Desire
Infatuation
Joy
Cheerfulness
Bliss
Delight
Happiness
Elation
Satisfaction
Ecstasy
Euphoria
Zest
Zeal
Excitement
Thrill
Exhilaration
Contentment
Pleasure
Pride
Hope
Aggravation
Annoyance
Grouchiness
Grumpiness
Exasperation
Frustration
Rage
Hostility
Bitterness
Hate
Loathing
Scorn
Spite
Resentment
Disgust
Revulsion
Contempt
Envy
Jealousy
Sadness
Agony
Anguish
Depression
Grief
Sorrow
Woe
Misery
Melancholy
Remorse
Loneliness
Rejection
Defeat
Dejection
Insecurity
Embarrassment
Humiliation
Fear
Horror
Shock
Terror
Panic
Nervousness
Anxiety
Apprehension
Worry
Dread

By comparison, no one has even attempted to make a list of a human being's fundamental beliefs, which are generated by a brain that is far more complex than what we find in most other living creatures. In all likelihood, it is our neural complexity that allows for our wide diversity of emotions; and I also believe that this huge array of emotions in turn contributes to many of the unique qualities we attribute to human nature. However, there may be some competition from a few of our mammalian cousins. Dolphins, for example, have a massive paralimbic lobe, a structure that human beings do not possess. This area is associated with the capacity for elaborate social communication and emotions relating to maternal feelings and separation anxiety. Thus, according to Jaak Panksepp, Distinguished

Research Professor of Psychobiology at Bowling Green State University, "dolphins may have social thoughts and feeling that we only vaguely imagine."[9]

In simpler animals, emotional responses become more and more limited. Most researchers, for example, limit nonhuman mammalian emotions to anger, fear, loneliness, and joy; and among reptilian species, emotions seem limited to primitive fight-or-flight reactions.

How Society Shapes our Beliefs

Emotions not only help us to maintain our beliefs but also defend us against other beliefs that threaten our worldview. When someone comes along with a different belief, what do we usually do? First, we dismiss him or her. After all, our brain has already done a lot of work establishing what we should and should not believe in, and the neural circuits have been set (neurons that fire together wire together, or so we currently believe). Besides, we're more likely to trust our own instincts over anyone else's. If our dismissal of the opposing opinion doesn't work, and the other person continues to press his or her point, we're likely to argue, with each side attempting to convince the other. Rarely, however, does anything change, and it doesn't matter how mature or immature your opponent may be. Try arguing with a four-year-old about the benefits of eating peas, and you'll see what I mean—all the logic, reason, and threats you can dream of won't convince kids that vegetables are more enjoyable than sweets. (Of course, there are exceptions to this rule. My own parents, for example, were successful in getting me to eat my peas. I'll explain how that happened in Chapter 4.)

> Oh, how sweet it is to hear one's own convictions from another's lips.
>
> —Goethe (1749–1832)

Perhaps it all comes down to conditioning, for, as I mentioned earlier, newborns have little choice but to accept basic beliefs given to them about the world. The recent discovery of mirror neurons also helps to explain why our brains are prone to absorb the behaviors and beliefs of others. The expression "Monkey see, monkey do" turns out to be neurologically true. When we see someone performing an action, whether peeling a banana or yawning, certain parts of our brain respond as if we were doing the action ourselves. (In the case of yawning, it's hard not to mirror the behavior as well. In fact, it's hard not to react with a yawn even when you simply read the word.) Although we are not aware of it, we are constantly monitoring and mirroring the behavior of our friends, the language of our parents, and the beliefs of the communities in which we live.

I would argue that much of human communication is primarily concerned with getting other people to think, believe, and behave as we do and vice versa. Our adoption of the beliefs of those nearest us helps us survive, primarily because it provides group cohesion. Without social consensus, we'd have anarchy and chaos, a perfect environment into which a dictator can step and impose his own set of beliefs and rules.

Group consensus, then, becomes an essential part of the belief-making process that integrates perceptual experiences, emotional values, and cognitive abstractions into a socially acceptable whole. How then do we sidestep arguments and fights when opposing belief systems collide, which happens, for example, when territorial boundaries of different cultural groups overlap? The answer, based on what we now know about the biology of belief, is that it takes a lot of hard work to build tolerance, acceptance, and appreciation, which are unfortunately uncommon in many corners of the world. It requires a conscious constant attempt to apply the spiritual and humanistic ideals on which society was based.

The Man Who Mistook His Belief for a Fact

As we go through life, we have to account for beliefs that appear bizarre or "abnormal." Different researchers analyze these oddities in various ways. Learning theorists focus on the roles of language and memory in the establishment of childhood beliefs; psychopharmacologists study how certain attitudes are affected by various drugs; and neurologists examine how brain lesions and strokes affect a person's assessment of the world. For example, a disorder known as Capgras syndrome can leave some victims convinced that the person standing in front of them is impersonating their doctor, child, or spouse; a patient may even believe that his own image in a mirror belongs to someone else. Patients with Cotard's syndrome actually believe that they are dead. A stroke victim may sometimes think that an arm or leg is missing when it is not. Such lesions and neurological disorders distort the perceptual processes of the brain, and this distortion forces the cognitive processes to come to bizarre conclusions about the reality of a situation. Consciously, such patients have little choice other than to accept the new reality they perceive because they have to believe in what they feel and see. To try to do otherwise could make them feel crazy.

In a case made famous by the neurologist Oliver Sacks, of Albert Einstein College of Medicine, a man who was shown a picture of the Sahara desert insisted that he saw a river with a little guesthouse, with people dining on a terrace under colored parasols. When this patient got up to leave Sacks's office, he "reached out his hand and took hold of his wife's head, tried to lift it off, to put it on. He had apparently mistaken his wife for a hat!"[10] The patient himself, however, did not think there was anything wrong with him. To maintain a sense of order and sanity in his life, he had to believe that his perceptions were real.

Actually, we all experience similar perceptual problems when we dream, but fortunately the brain paralyzes our body during dreaming so that we cannot physically respond. Otherwise, we'd thrash, yell, and run around in our sleep. Dreams seem unreal only when we awake and a different system of belief—and reality—takes over. But psychotic individuals will have more difficulty distinguishing between fantasies and facts, in part because the dreaming mechanism of their brains is unable to shut down during the waking cycle of the day.

The study of dreams, dementia, and psychoses can help us understand how strange beliefs are formed, but as I and many of my colleagues in psychiatry have often reflected, some of the visions reported to us by schizophrenic patients seem to have a certain degree of value and truth. Since my research suggests that the brain may be neurologically biased toward perceiving or generating spiritual imagery, the visions themselves should not be considered evidence of psychosis; rather, it is the behavior of the patient that needs to be evaluated. I know of several cases in which the religious imagery generated in a psychotic state inspired the patients to recover. For example, one young woman who was addicted to methamphetamines began to hear angelic voices that told her to stop using drugs and to leave the abusive relationship she was in. According to the therapist who treated her, the voices were clearly hallucinatory, but they still held the power to heal.

How do we decide if such visions are merely hallucinatory or if they might possibly represent some form of spiritual or metaphoric truth? We can't, because a false belief is determined by many converging factors, and in a clinical setting one of those factors involves the belief system of the person who is treating the patient. Thus, what is normal and what is abnormal are also beliefs that are consensually agreed on among members of a cultural or professional group.

In western psychology, there is little consensus supporting a belief in divine intervention, but other groups and cultures believe differently. That is why, historically, in various parts of Asia and the Middle East, people we might consider psychotic would instead be thought of as "touched by the hand of God," and as simply unable to integrate divine power into their everyday lives. Instead of being ostracized, as so often happens in America, these mentally ill individuals have been treated with compassion.

The Mathematical "Unprovability" of Truth

With regard to those beliefs we hold most sacred—whether they are about politics or religion or love—we tend to spend a lot of time convincing ourselves, and others, that they are valid. And the fundamental tool we use is logic. We search for consistency and coherence, and if the numbers don't add up, we go back to the lab and try again. People may never come to an agreement concerning morals or evolution or the existence or nature of God, but nearly everyone concurs that $1 + 1 = 2$. Mathematics uses symbolic logic, and as long as the symbols are clearly defined, mathematical proofs are a gold standard for establishing what is true.

In the early twentieth century, it was generally accepted that everything real could be substantiated by mathematics, and that everything proved by mathematics was true. However, in 1931, the brilliant mathematician Kurt Gödel created an elegant though rather nasty-looking formula to challenge that belief. Here's a tiny segment of his equation:

$$P(x0 \ldots xn) \circ (\$n,d)\{S([n]d + 1, x2 \ldots xn) \;\&\; (k)$$
$$[k < x1 \circ T([n]\, 1 + d(k + 2), k, [n]\, 1 + d\,(k + 1), x2 \ldots xn)]$$
$$\&\; x0 = [n]\, 1 + d\,(x1 + 1)\}$$

With this equation, Gödel demonstrated that any mathematical or symbolic system of logic will always be incomplete and contain assumptions that cannot be proved.[11] This means that nearly every scientific notion we hold will contain suppositions that may be false. Truth cannot be entirely known, for no matter how much evidence you collect, your knowledge will always be incomplete.

Douglas Hofstadter, a professor of cognitive and computer sciences at Indiana University, believes that Gödel's incompleteness theorem directly applies to our beliefs about who we are:

> Just as we cannot see our faces with our own eyes, is it not inconceivable to expect that we cannot mirror our complete mental structures in the symbols which carry them out? All the limitative theorems of mathematics and the theory of computation suggest that once the ability to represent your own structure has reached a certain critical point, that is the kiss of death: it guarantees that you can never represent yourself totally.[12]

This suggests to me that faith will always play an essential role in human life, allowing us to trust our beliefs so that we can survive and glean meaning and value from the world. If we didn't trust our beliefs, we might end up living in perpetual doubt; the amount of stress hormones our brains would secrete under such conditions could physically atrophy the brain.[13] We all must live between the extremes of absolute doubt and certainty, trusting that our beliefs bear some semblance to the nature of reality and truth. But, by acknowledging that our beliefs are, at most, a "best guess," we can stay open to other opinions and views.

Gödel and the Liar's Paradox

Can we ever ascertain the truth of a single statement? This question has been pondered since the days of Socrates, who purportedly said, "The one thing that I know is that I know nothing"—a paradox, for if Socrates knows nothing, he cannot know that this statement is true.

People tend to make certain logical assumptions about their thoughts. First, we assume that some statements are true (in philosophy this is called the law of identity). Second, we often assume that no statement can be both true and false (the law of contradiction). Third, we often assume, mistakenly, that every statement must be either true or false (the law of the excluded middle). If a friend tells you that it is raining, he is either lying or telling the truth. Or is he? After all, it is probably raining somewhere in the world at that moment.

But what happens if a man comes up to you and says, "I am a liar"? If the statement is true, then he's not a liar, so the statement is false. But the statement can't be false, because if that were the case, he would be telling the truth and would not be a liar. The statement is an example of an unresolvable contradiction, and thus is neither true nor false. "This painting is beautiful" is another example of a statement that could be true for one person and not another. Neurologically, truth and fiction are subjective values created by the brain. As far as human survival is concerned, it isn't necessary to know what is absolutely true; you simply have to use logic, as the following riddle demonstrates. It's called the Liar's Paradox, and Gödel used a version of it to illustrate his incompleteness theorem.

Imagine that you are journeying down a path and you come to a fork in the road. There you meet two people, and a sign that reads:

> **One of these paths will lead you to safety, but the other will lead to death. The two men in front of you both know which path is which. However, one person always lies and the other always tells the truth. You may ask only one question in order to decide which path to take.**

If you could ask two questions, the problem would be simple. You could ask one person how many eyes you have. You would immediately know who was telling the truth and could then ask that person which path to take. Unfortunately, you get only one question.

Here's the solution. Ask either man which path the other person would tell you to take to reach safety, and then take the opposite path. Assume, for a moment, that path A leads to safety and path B to death. The liar will tell you that the truth teller would say to take path B. The truth teller will tell you that the liar would say to take path B. Obviously, you should take path A. Gödel tried to use similar logic to prove the existence of God, but he overlooked his own theorem, which implies that every equation contains assumptions that might be false. In the example above, you probably assumed that the sign was stating the truth. What if it had been written by a liar?

Is There a Better Way to Ascertain the Truth?

Every belief begins with a different assumption, and every assumption leads to different conclusions and truths. Western religions often begin with the assumption that God is the ultimate creator, and thus that the world and human beings emerge from God. Evolutionary psychology turns this assumption upside down: first the universe is created from a cosmological big bang, from which life emerges. Eventually human beings appear; and from the minds of humans, the notion of God is born.

In fact, every religion and every philosopher usually begins with a basic unquestioned assumption. Descartes, for example, began by doubting the truth of all his previous beliefs. This doubting led him to the conclusion that though he might doubt everything else, he could not doubt that he himself existed. "I think, therefore I am." He maintained a belief that the body and thoughts were separate entities. But Descartes' contemporary Spinoza regarded thoughts and the physical universe as different aspects of a single substance, which he alternately called "God" or "nature." In contrast, certain forms of Buddhism reject all these beliefs, suggesting instead that the world, as the mind perceives it, is an illusion. Buddhist practices of meditation attempt to silence all thought—all of one's beliefs—so that the true nature of reality can be perceived.

Depending on the assumptions we begin with, our beliefs about reality will differ. Which belief system, then, captures the most accurate view of the world? Gödel's theorem suggests that we can never know for sure, but Spinoza offered another solution that I find particularly intriguing. He believed that truth consists of three kinds of knowledge: opinion, reason, and intuition. Opinion, which he saw as a basic form of belief, is based on a combination of sensory experience, imagination, and a partial assemblage of ideas—a concept that anticipated our current understanding of the perceptual and cognitive processes of the brain. However, Spinoza held that reason constitutes a more comprehensive form of belief because it applies rules of logic. This idea, too, anticipated our neurological model of conscious beliefs.

For Spinoza, though, the highest form of knowledge was intuition, which takes the individual beyond personal beliefs and brings him or her closest to reality and truth. Spinoza believed that the evolution of one's thinking from opinion to reason to intuition brought with it a deep sense of peace and happiness, and freedom from anxiety, fear, and despair. In this state, one begins to experience the essence of an infinite, indivisible "substance,"[14] a term that Spinoza used to simultaneously embrace God, nature, and the sum total of reality itself:

> By God, I mean a being absolutely infinite—that is, a substance consisting in infinite attributes, of which each expresses eternal and infinite essentiality.[15]

Spinoza managed to abolish the dualism between mind and nature, but in the process, he removed the "otherness" of a personal God who could intervene in human life. To the religious orthodoxy of the seventeenth century, such a nonpersonal god could never provide a sense of comfort, meaning, or solace, and so Spinoza was branded an atheist by his contemporaries. Today, however, many people have no difficulty in embracing this form of natural spirituality.

Spinoza's notion of intuition captures my interest because it correlates with the way our brains create a holistic image of the world by putting all the pieces together to create something greater than the parts. Intuition allows us to comprehend what the senses cannot perceive; and as far as my research into the neural mechanisms of spirituality suggests, we can enter into intuitive states through the act of meditation and prayer. These processes can enhance our lives by allowing us to circumvent the conceptual errors embedded in logic, reason, or personal opinion. Intuition, creativity, and spiritual practice may all provide better means for apprehending reality and truth more accurately. Throughout this book, I will return to this astonishing but scientifically plausible hypothesis.

Descartes' and Spinoza's Legacy

Antonio Damasio, a professor of neurology at the University of Iowa School of Medicine, argues that Descartes erred by assuming that the mind and body were independent of one another and that human emotions and rationality were basically opposed to each other. Descartes argued in favor of reason over emotion, but Damasio contends that our emotions are fundamental to our ability to make decisions and understand the world, a view that is now widely accepted in the neurosciences.[16]

Damasio, like Plato, Socrates, and Descartes, tries to make a distinction between knowledge and belief, arguing that knowledge is the direct perception of information about the world, whereas belief is the qualification we place on the certainty, accuracy, or truth of that perception.[17] The only problem with this argument, as I see it, is that every neural process, including the gathering of sensory information, alters our perception of the world.

Damasio considers beliefs a composite of memories and internal emotional states, a creative process of the mind that can, from moment to moment, use, discard, or modify our intuitions about the world. As Damasio points out, "Our neural and cognitive systems allow us to jump to a conclusion or even to an action without relying on intervening cognitive steps."[18] This view clearly leans more toward Spinoza than toward Descartes, for here Damasio is integrating intuition, feelings, and reason into a holistic worldview. He defines spirituality as "an organizing scheme behind a life that is well-balanced, well-tempered, and well-intended";[19] Damasio thus believes that the neurobiology of religious experience—such as blessedness, beatitude, and grace—will eventually be mapped out in the lab.

The Flexibility of Beliefs

Expanding on Damasio's model concerning knowledge and belief, Howard Eichenbaum of Boston University and J. Alexander Bodkin of McLean Hospital suggest that knowledge is more flexible than belief: "Knowledge is a disposition to behave that is constantly subject to corrective modification and updating by experience, while belief is a disposition to behave that is resistant to correction by experience."[20] This line of thinking, however, ignores the wealth of cognitive research demonstrating that belief systems can be flexible, and that they can rapidly change—especially during childhood and adolescent stages of development—without the person's conscious awareness that his or her assumptions have been altered.

I propose that beliefs are always in flux, and that the human brain is continually imagining and intuiting alternative perspectives on reality. This flexibility may have evolved to allow the brain to adapt its thinking to the new and unusual situations it encountered.

The brain is also adept at imagining potential realities. For example, around 400 B.C., Democritus envisioned an atom, which he considered the fundamental building block of the universe. In the late 1800s, Joseph J. Thomson proposed the "plum pudding" model, suggesting that there were even smaller parts within atoms. Other scientists of his time proposed the "strong force" and "cloud" models of molecular interactions that later became incorporated into the theories of quantum mechanics. One person builds on the ideas of another. This chain of ideas led to the atomic bomb as well as nuclear energy, nuclear disarmament treaties, and even global-warming scenarios. They all originally arose from the brain's neural capacity to generate flexible beliefs.

Our beliefs, therefore, are an assemblage of perceptual experiences, emotional evaluations, and cognitive abstractions that are blended with fantasy, imagination, and intuitive speculation. In spite of our lapses of memory, our inconsistencies of logic, and the inherent shortcomings of consciousness, humans have done a pretty good job at surviving. For better or worse, we reinvent the world every day, searching for the ultimate reality we call truth, enlightenment, or God.

No other organism seems to demonstrate this passion for truth. Perhaps this is due to the nature of consciousness itself, which allows us to reflect on the beliefs we construct about the world. "Other species have a very limited ability to reflect, and their sensitivities are channeled down rather narrow sets of possibilities," writes Daniel Dennett. "We, in contrast, are *believe-alls.* There is no limit, apparently, to what we can believe, and to what we can distinguish in belief."[21]

The sleeping lion is not concerned with such matters. When threatened, it roars, runs, attacks, and then returns to its repose. The lizard seeks out food when it is hungry, then buries itself in the sand. But human beings seem to be in a perpetual state of alertness, always aware that things are not exactly as they seem. We find evolutionary processes at work and discover the secrets of cloning; then we turn on the news, bombarding ourselves with images of disaster. We're never certain if we have enough money, enough love, or enough security, and so we do not sleep as well as the lion. We take vitamins and drugs. We think. We read and study and go to religious services, seeking answers to our questions, and seeking truth. If we find discrepancies, we may change our beliefs, but no matter how we revise the map, some new piece of information is bound to shake us up.

You can call it human nature or fate, but in such a state of perpetual uncertainty, how do we find happiness and peace? The first step is to learn that we do not need to grasp the absolute truth in order to survive. We can appreciate the mysteries of the universe and the mysteries of the brain; and we can learn to trust our intuition, to have faith in our biological and even our spiritual drives. When we do so, we will find that it is easier to sleep at night.

Chapter 3

Reality, Illusions, and the Aunt Who Cried Wolf: The Construction of Perceptual Beliefs

> What can we know? What are we all? Poor silly half-brained things peering out at the infinite, with the aspirations of angels and the instincts of beasts.
>
> —Stark Munro, a fictional character created by Sir Arthur Conan Doyle

ONE MORNING, MANY YEARS AGO, MY AUNT ROSE WOKE UP to an astonishing sight outside her window. There, sitting on the lawn, sunning itself in a small patch of early morning light, was a rather large orange-colored creature. "Oh!" she exclaimed, "There is a wolf in my backyard!" My aunt lived in a city—Philadelphia—and the likelihood of a wolf being found within 100 miles of her home was extremely doubtful. Still, she called the police. "That's right, officer," she said to the incredulous policeman, "There's a giant wolf in my yard."

A car was dispatched to her house, and on arrival, the officer sarcastically asked, "All right ma'am, where's the wolf?" "There," my aunt pointed, as a furry creature scampered over the fence. "Lady, that was no wolf," the policeman said, "It was just an alley cat." She thanked the officer and off he went; but when her neighbor came over to ask what all the commotion was about, she replied, "Oh, nothing, really—we just had a wolf in our backyard."

For my aunt, the wolf was reality. That's what she saw, and no amount of evidence was going to convince her otherwise. Some people might question her sanity, or suggest a visit to an optometrist; but this type of error is more common than you might think, for people mistakenly identify wild animals in their yards all the time. Because first impressions make a strong emotional impact, my aunt couldn't get the idea out of her head that she had seen a wolf. Fear of embarrassment might have also strengthened her resolve to cling to her belief.

On the other hand, how do we know that my aunt didn't see a wolf? It may sound improbable, but other people have found stranger creatures lurking in their yards. In Oregon, a woman who came out to get her morning paper stumbled over a three-foot alligator, the missing pet of a neighbor who had been distracted while caring for his twelve-foot Burmese python. After all, you and I are making lots of assumptions about my aunt's situation; for example, that wolves don't roam in downtown Philadelphia, and that a policeman's observation will be more accurate than an elderly woman's. What if a truck full of army officers had come along and they had also seen a wolf? Then whom would you believe, the army officers or the police? It's even possible that the policeman did see a wolf, but his skepticism precluded that possibility, for if the data do not fit our version of reality, then we are likely to dismiss the data in favor of a more consistent view.

For that matter, how do you know that I'm telling you the truth about the policeman and my aunt? I wouldn't be the first author to fabricate a tale, knowing that people tend to believe what they read, especially if the writer has impressive credentials. How do you know if anyone is telling you the truth about anything? Think of a "brain in a vat" scenario in which a false sense of reality is fed into the minds of unsuspecting people by an evil scientist—or in the case of the movie *The Matrix,* by a maniacal computer. This is also a metaphor used by cognitive theorists in discussing the nature of human consciousness, and a useful description of what political or religious fanatics may attempt to do with prisoners of war. In many ways the metaphor of a brain in a vat has scientific validity because we are all biologically confined to the limitations of our physical senses when we try to assess reality.

So what is the truth about reality? Do objects really exist "out there," or is everything an illusion, an ephemeral blip of consciousness floating in a brain that is suspended within an infinite universe, created and directed by the quantum complexities of physics? Such questions have been debated for centuries, and still no consensus exists on what reality actually is. We could just as easily substitute the seventeenth-century British philosophers John Locke and George Berkeley for the policeman and my aunt. Locke and the policeman would probably argue that the wolf was a fantasy, since there was visual evidence that the creature was merely a cat. But Berkeley would argue that reality itself doesn't exist, that all things are immaterial. "To be is to be perceived," he would say, and thus he might be more receptive to my aunt's point of view. In today's scientific community, the majority would probably lean toward Locke's view and assume that an objective reality exists out there. They'd humor my aunt rather than believe her.

Reality, Philosophy, and God

George Berkeley

John Locke

The question whether or not there is a reality "out there," or if we can ever actually know what reality is, has puzzled philosophers throughout history. Socrates, for example, argued that the senses do not grasp reality in any way. The Buddha believed that the reality we perceive is nothing more than an illusion that can be pierced only through deep contemplative practice. John Locke (1632–1704), however, believed that the world "out there" actually existed, but that human beings had to rely on their limited senses to experience it, an empiricist view that continues to dominate the natural sciences today. For Locke, objects exuded qualities (motion, solidity, etc.) that left impressions (color, taste, sound, etc.) on our minds. However, these impressions were nothing more than ideas (red, sweet, loud, etc.), and thus our experience of the world could be only an internal representation of the world "out there." Thus, he concludes, we can never really know with certainty what qualities any object, including God, may have. We can be certain only of our ideas and beliefs. Our current understanding of the brain generally supports Locke's premises.

George Berkeley (1685–1753), an Anglican bishop, took exception to Locke. Since no connection could be made between ideas and the objects they are supposed to represent, Berkeley feared that such a perspective would lead to theological and religious skepticism, particularly since an immaterial God could not be, at least by Locke's definition, perceived by our senses. So Berkeley took a different approach, arguing that only the ideas we directly perceive are real. Furthermore, you cannot conceive of an object that exists independently of the mind. Since matter doesn't actually exist, the only true reality was a spiritual one. In this sense, Berkeley saw God's spirit and the human spirit as one and the same.

In western philosophy, questions concerning the nature of reality have rarely been separated from issues about God. Even in the field of modern astrophysics, God continues to be debated: "With the success of scientific theories in describing events, most people have come to believe that God allows the universe to evolve according to a set of laws and does not intervene in the universe to break these laws," writes Stephen Hawking.* "However, the laws do not tell us what the universe should have looked like when it started—it would still be up to God to wind up the clockwork and choose how to start it off. So long as the universe had a beginning, we could suppose it

* S. Hawking. 1988. *A Brief History of Time*. New York: Bantam, p. 146.

had a creator. But if the universe is really completely self-contained, having no boundaries or edge, it would have neither beginning nor end: it would simply be. What place, then, for a creator?"

From a neuroscientific perspective, since the brain is capable of perceiving transcendent realities, does this mean that such realities exist? Berkeley says yes, Locke says we'll never know, but Hawking would probably say "Maybe yes, maybe no." No matter how you look at it, science does not preclude the existence of spiritual realms.

The Boy Who Cried "God"

Suppose we were to change my aunt's story by substituting the word "God" for "wolf"? Imagine that your neighbor—say, for example, a ten-year-old boy—came to you and told you that he had just seen God in his backyard. Would you say that his apparent experience resulted simply from a child's overly active imagination? Probably. But what if your neighbor was the pope, or the Dalai Lama? Would you then give more credence to the story and assume that he may have indeed perceived some spiritual presence? If your answer is yes, then philosophically, you—and millions of others—will have sided with Berkeley's notion of reality, according to which the mind is capable of perceiving immaterial and spiritual dimensions. "God exists," Berkely might say, "because God has been perceived."

The Illusions of Reality

The world of contemporary neuroscience integrates Locke and Berkeley by arguing that two realities exist simultaneously: the objective world in which we live, and the subjective world constructed by the brain. How and where the subjective and objective realities intersect within our brain remains a mystery.

In the 1970s, cognitive psychologists and neurologists became interested in optical illusions because these phenomena provided insight into the mechanisms of perception that shape our views of reality. They discovered that the brain often creates images that do not exist in the world. The following illustrations help to demonstrate this. In Exhibit 3-1, everyone sees a white square, even though the drawing shows only the four black incomplete circles. Our eyes don't construct complete circles, but they do construct imaginary lines to complete the image of the square. In addition, most people perceive the square as brighter than the surrounding white areas of the page. In reality, there is no square, and there is no difference in brightness.

Exhibit 3-1

What is happening in the brain? The visual centers are designed to recognize patterns, and if a particular form is vague (vagueness can be caused by high or low contrasts in light, rapid movements, competing shapes, etc.), the brain will attempt to clarify the ambiguity by making a guess about what the object might be. In Exhibit 3-1, your perceptual mechanisms have been tricked into constructing invisible lines. Other processes in different parts of the brain create the sense of interior and exterior space and assign values of contrast that are used to maintain the visual sense of a square. Without your awareness, the brain also decides to focus less and less on the black areas, and the newly constructed image involving the white space is then held in working memory, which fixes the image in your consciousness.

Exhibit 3-2 is one of the earliest optical illusions, which fascinated the ancient Greeks because it forces your brain to transform a two-dimensional drawing into a three-dimensional cube. According to the psychologist Roger M. Shepard of Stanford:

> Our perceptual machinery for [identifying] objects in three-dimensional space is deeply entrenched in our nervous system and wholly automatic in its operation. Without our bidding or

even our awareness of its existence, this machinery immediately goes to work on any visual input, including the visual input provided by a two-dimensional drawing. As a result, we cannot choose to see a drawing merely as what it is—a pattern of lines on a flat, two-dimensional surface.[1]

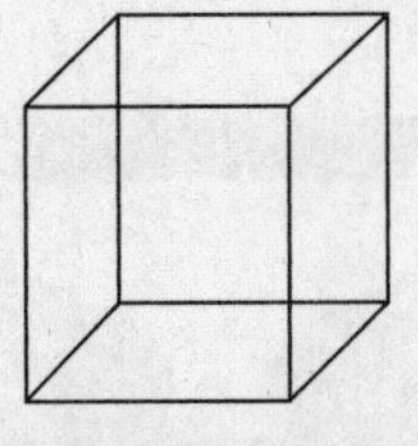

Exhibit 3-2

When you look at the cube in Exhibit 3-2, you will notice another perceptual phenomenon: the orientation of the cube will reverse, making the surface that appears as the front face flip suddenly and become the back face. However, your visual processing will not allow you to see both shapes at the same time. Furthermore, a process known as perceptual bias will make most people prefer one perspective over the other.

Perceptual reversals are not limited to vision; they can be experienced with any of our senses. Repeated words, if listened to on an endless tape loop, will cause auditory reversals,[2] and water at room temperature will feel either hot or cold, depending on the temperature of your hand. When I was young, my friends and I would make use of this phenomenon to trick each other. We would blindfold someone and then tell him that we were going to stick his hand into boiling water. The blindfolded kid, of course, didn't really believe us, but when we stuck his hand into ice-cold water, he'd yelp and pull his arm away. This is not a perceptual distortion of reality but an interpretive distortion: the mind thinks that one thing is happening but experiences something different, or unexpected. The reaction is usually shock. The same thing happens if you bite into a piece of food thinking it is one thing, when it's really something else: your first reaction will be disgust because the flavor doesn't match the memory associated with the idea.

The classic vase-face illusion (Exhibit 3-3) demonstrates how the brain turns ambiguous shapes into recognizable form; it also shows how perceptual reversals and biases occur. This illusion has helped neuroscientists to identify other reality-construction mechanisms of the brain, including pattern, object, and facial-recognition circuits.[3]

Exhibit 3-3

According to Al Seckel, a research fellow at the California Institute of Technology, "the visual system represents or encodes objects primarily in terms of their contours . . . [and] the sudden reversal that you perceive may be due to your shift of attention on the shape of the contour."[4] As with the cube illusion, the brain cannot see both images at once. Furthermore, it will tend to prefer one image over the other.

The vase illusion also demonstrates another important principle of perception: the brain takes only a few elements from what it sees to construct an internal image. It's an efficient mechanism, but one that leaves us vulnerable to misinterpretation. In addition, the more ambiguous a figure is, the greater the possibility of error. For example, if you turn Exhibit 3-3 upside down, you'll still see the vase—or perhaps a candlestick—but you'll probably find that the face all but disappears. It's contours, which are vague to start with, are now seen out of context (human faces are rarely seen upside down).

Sometimes our visual preferences are so strong that we can see only one of the images in an optical illusion. What, for example, do you see in Exhibit 3-4, which has been a popular illusion since the 1800s?

Exhibit 3-4

Most people can find one image but not the other, even after they've been told to look for a young woman and an old hag.* In this illusion, the significant ambiguity is the ear-eye; for once the brain decides what it sees, it precludes the other choice. In a similar way, once my aunt saw a wolf, that image became fixed, and she could not see the contours of a cat.

Seeing is Believing

Holistic grouping processes play an essential role in the brain's construction of reality by gathering together a few perceptual elements, intuiting or imagining others, and comparing its construction with stored memories. Information that doesn't fit is often excluded from both memory and consciousness. The same thing occurs when we see shadows at night or an unusual movement in foliage. First, the brain will try to ascertain whether the stimulus is alive or not—a survival mechanism. If we think we see a dark shape in the forest, the impulse is to run away. If we were mistaken, well, no harm done. It's a normal human instinct to be frightened by uncertainty.

* The young woman is facing away, toward the left, with her long hair flowing down her back. To visualize the hag, see the young woman's chin as a giant nose, while her own petite nose becomes a wart. Her necklace becomes the hag's mouth, and her ear becomes the eye of the old woman. For some people, the hag is more difficult to see because the features are anatomically distorted.

Ambiguous shapes and contrasts also create false boundaries in the brain; and, since the human brain also has specialized neurons that are designed to identify faces, we tend to look for such images when we are confronted with indistinct forms. Most cultures even make a game of it: Americans see a man in the moon, the Chinese see a rabbit, the Pima Indians see a coyote, and the ancient Mayans found goddesses. The Greeks outlined mythological forms in the stars, and children find dragons in the clouds. And of course seekers of miracles can see angels in reflections of light. If you happen to live in a community that believes in demons and ghosts, then you'll find ample opportunity in the shadows to support that cultural view.

Given the proper conditions of light, nearly anyone can be tricked into seeing objects that do not actually exist, or into not seeing objects that do exist. In Exhibit 3-5, the contrast between the black and white areas gives the brain the illusion of moving gray circles that flash on and off in the white areas. However, if you stare at any specific white space, no gray circle appears.[5]

Exhibit 3-5

The same thing occurs if we think we see someone or something out of the corner of our eye, but when we look directly, we see nothing. In mythology, this type of perception has been used to explain the presence of fairies, goblins, gnomes, and other unearthly creatures, even the figure of death. Now you see them, now you don't.

Differences in contrast can also trick the mind into distorting parallel lines, as Exhibit 3-6 demonstrates.[6] The lines neither widen nor tilt, but you virtually have to use a ruler to convince yourself that they're straight. These types of illusions have been used to identify how images are processed by the retinal structures of the eye. The illusion may be caused by orientation-sensitive cells in the part of the brain that processes vision.[7]

Exhibit 3-6

Other optical illusions can give the impression of movement, and some of the best examples of this visual phenomenon were created by Akiyoshi Kitaoka, of Ritsumeikan University, whose research has been used to explain not only how movement is constructed in the brain, but also why the brain is prone to creating phantom images.* Perceptual scientists refer to this as the binding problem: the brain attempts to form a coherent perception of the world by putting together information (shapes, colors, movements, etc.) in ways that are consistent with prior experiences and memories.

What I want to emphasize is that everything we see is an illusion, in the sense that our eyes, memories, and consciousness can envision only a symbolic representation of the world. The illusions I have discussed in this chapter exemplify how our brain converts reality into lines and shapes, with depth and color, constructing something that differs from actual appearances in the external world. Light may exist, but light is nothing more than a range of electromagnetic fre-

* You can view these illusions at http://www.ritsumei.ac.jp/~akitaoka/index-e.html.

quencies that stimulate neural mechanisms in our eyes. The color we see is a label—call it blue, green, or fuchsia—an arbitrary symbol used to describe a particular experience. Some animals may find it useful for survival, but a cave-dwelling fish has no need for that type of neuronal circuitry.

From a neurological perspective, we might argue that color does not actually exist anywhere, except in the deepest recesses of the brain. What we "see" is a patch of neuronal activity, a neurochemical interpretation of light patterns that enter our eyes and are transmitted to the visual parts of the brain, where they are broken down and processed, memorized, and recalled as needed. For that matter, we don't even need our eyes to see. With a little technological help, we can use the tongue instead, as Dr. Paul Bach-y-Rita has done with people who are blind.[8] With the aid of a camera and a device that translates incoming light into tactile impulses that can be felt inside the mouth, a person's brain can translate the stimuli into images that are similar to what you and I perceive. When Marie-Laurie Martin, who had been blind for thirty-nine years, used the device to view a candle flame, she was surprised at how small it was, and how it flickered. Previously, she had imagined that it was a big ball of fire that didn't shimmer or move.[9]

The brain, with its complex functions, can process sensory stimuli in any number of ways. A neurological condition known as synesthesia even allows some individuals to see music, hear shapes, or taste the flavor of a name. Such people are not crazy; their brains are simply processing sensory information in unique ways. Their experience of reality is no less valid than anyone else's and may even provide an enriched experience of life.

Just as music is a neural interpretation of sound, color is a neural interpretation of light. The eye and visual system in the brain help differentiate various light frequencies. We see not the frequency but the color. We use language to communicate about color as a concept. If someone doesn't come along and teach us that a particular frequency has a name (e.g., "red"), we might not consciously look at it as a particular color. In fact, some cultures divide and label the color spectrum in different ways, and the members of those communities cannot recognize certain colors that people in other communities identify.

Members of the Berinmo tribe of Papua New Guinea make no distinction between blue and green, but they can be taught to distinguish these colors. This supports the theory that color categorization is a language-bound, higher-order process in the human brain.[10]

To the human brain, color is primarily a subjective experience. The experience differs from person to person and culture to culture. Trevor Lamb and Janine Bourriau, professors at Cambridge and Oxford respectively, write:

> Although the idea of "colour" may seem a simple concept, it conjures up very different ideas for each of us. To the physicist, colour is determined by the wavelength of light. To the physiologist and psychologist, our perception of colour involves neural responses in the eye and the brain, and is subject to the limitations of our nervous system. To the naturalist, colour is not only a thing of beauty but also a determinant of survival in nature. To the social historian and linguist, our understanding and interpretation of colour are inextricably linked to our own culture. To the art historian, the development of colour in painting can be traced both in artistic and technological terms. And for the painter, colour provides a means of expressing feelings and the intangible, making possible the creation of a work of art.[11]

I'd like to bring up one more visual oddity, before we move on to sound. Although we have neural receptors for various colors, including those that respond to "blackness" and "whiteness," there is no neural receptor that distinguishes any gradation of gray. No one knows for sure where the experience of gray occurs, but one theory suggests that it is a concept fabricated in another part of the brain when both the blackness and the whiteness receptors are turned off.[12] Gray, like many other colors we can imagine, is a belief construction within the brain—a form of understanding, a thought.

Is It Live, or Is It Memorex?

As with vision, the auditory mechanisms of the brain select a narrow band of experience from all the noise that invades the ear. And just as the eye transforms light into a limited range of identifiable elements, the ear discards much of what it picks up from the environment and fills in the auditory gaps with perceptual guesswork and imagination. The auditory part of the brain is also prone to error. For example, repetitive patterns of random sounds can be transformed into words, complete sentences, even music. Diana Deutsch, a professor of psychology at the University of California, San Diego, created an audio CD of auditory illusions, called *Phantom Words and Other Curiosities.** You'll hear words and phrases that trick the mind, including a repetitious phrase that turns into a song.

I listened to her CD with a group of colleagues, playing a track that rapidly repeated a two-syllable word. Each of us heard something different, and no one was able to identify the word that was actually spoken, which was "Harvey." One person heard "big fig," another heard "fake," and I heard "coffee." The moment I said that, everyone else also heard "coffee" (demonstrating the brain's propensity to bias itself in the direction of social consensus). Something even more unusual happened: when we moved about the room, we would hear different expressions and words. According to Dr. Deutsch, people will also hear words that relate to their current concerns and ethnic backgrounds:

> If they are on a diet, they may hear words that are related to food; if they have had a stressful day they may hear words that are related to stress; and so on. In fact, so strong is the influence of meaning on what is perceived, that people sometimes hear voices speaking in strange or unfamiliar accents, so as to create for themselves words and phrases that are particularly significant to them. . . . Native speakers of Chinese sometimes hear Chinese words, and native speakers of Spanish sometimes hear Spanish words. This impression can be so strong that people are sometimes convinced that such "foreign" words have been inserted into the tracks, though in reality this never happens. Also, the words that are heard often appear to be spoken by different voices, each of which has a distinctive quality.[13]

* You can hear some of these illusions on the Philomel Record website at http://www.philomel.com.

I suppose that if one was in the "right" environment—a church, perhaps, or a cemetery—and the "right" state of mind, the sounds and echoes could be construed as a heavenly voice, or even a ghost.

Jerome S. Bruner, a former professor at Harvard and Oxford, describes a variety of experiments that show the illusory nature of both sight and sound. For example, if you use a dimmer switch to control a light while a buzz saw is heard in the background, the sound of the saw will seem to rise and fall as you turn the brightness of the light up and down. And if you stare at a light while a faint tone is played and then turn off the light and sound, when you turn the light back on without the music you will hear a tone that isn't there.[14] In other experiments, different tones and repetitious sounds have caused a person to see two flashes of light when only one flash appeared.[15] Other studies have found that sound can evoke visual images that do not actually exist.[16] In fact, all of our senses play integral roles in providing a coherent and realistic perception of the world.

> Believe only half of what you see and nothing that you hear.
>
> —Dinah Mulock Craik, *a successful Victorian author who was the daughter of an evangelical preacher*

Perceiving God

For centuries, visual stimulation and auditory stimulation have been used to induce mystical experiences. In many religions, chanting, drumming, music, incense, candles, and colored lights can trigger visions of otherworldly realms. Gregorian chants, Hindu mantras, and Navaho sand paintings have all been used to transport the practitioner into states that alter the neural processing of his senses.

In the Pentecostal churches, some members practice a very unusual technique that they believe brings them into direct communication with God: they speak in tongues. To the casual listener, what they say sounds somewhat like a foreign language; but for the practitioners, this speaking is a door to an inner experience that is profoundly meaningful. Are they simply hearing unintended words, as described by Deutsch's auditory experiments? In my opinion, as I will explain in detail in Chapter 8, something much more complex is happening in the brain, for as Dr. Bruner emphasizes, people reshape experiences to fit their personal preferences "with a view toward meaning and signification, not toward the end of somehow 'preserving' the facts themselves."[17] Spiritual practices go one step further by deliberately disrupting the perceptual organization of the brain. In this way, practitioners alter the reality in which they normally live, in the hope of perceiving spiritual realms. In later chapters, I will discuss how prayer, meditation, and other spiritual practices can transport practitioners into extraordinary states of consciousness that our everyday perceptions exclude.

The Anatomy of the Invisible Man

Perhaps the most disturbing illusions are those that give us a tactile sense that something exists when we know it isn't there. Some amputees continue to experience the presence of the removed part (such as a limb) for weeks, sometimes years. They can experience pain in missing fingers and toes; and some people who experience a "phantom limb," as this phenomenon is referred to, even believe that they can hold and feel objects in a missing hand. In one case, reported by V. S. Ramachandran, director of the Center for Brain and Cognition at the University of California in San Diego, when he went to pick up a coffee cup that a patient claimed to be holding with a missing arm, the patient actually cried out in pain.[18]

Illusory sensations of movement can also be elicited in patients with damage to the right side of the brain.[19] This suggests that the two brain hemispheres process reality in different ways, and thus our belief systems depend on an integrated coherence between these two perceptions. When coherence is disrupted, a person may believe that a paralyzed limb is perfectly fine, or that it actually belongs to someone else. Such patients are not lying; they are simply conveying their sense of reality. Fortunately, such symptoms often disappear after a few weeks, along with the memory that they ever existed. Dr. Ramachandran explains:

The left hemisphere's job is to create a belief system or model and to fold new experiences into that belief system. If confronted with some new information that doesn't fit the model, it relies on Freudian defense mechanisms to deny, repress or confabulate—anything to preserve the status quo. The right hemisphere's strategy, on the other hand, is to play "Devil's Advocate," to question the status quo and look for global inconsistencies. When the anomalous information reaches a certain threshold, the right hemisphere decides that it is time to force a complete revision of the entire model and start from scratch.[20]

The memory of the sense of our bodies becomes so ingrained in the neural circuits governing self-experience that the brain has difficulty reorganizing itself after a crippling accident or stroke. Sometimes the reality of the situation is so painful that the person simply cannot accept the truth. In such cases, a false belief in an uninjured body can be constructed, triggering an emotional memory that feels utterly present and real. This may have been the case with a man who experienced phantom erections after his penis was removed.[21] It may also explain the underlying neural mechanisms of anorexia nervosa,[22] in which patients actually "see" fat on their bodies where there is none. Usually such a problem is viewed as a psychological delusion, which the American Psychiatric Association defines as a "false belief based upon an incorrect inference about external reality," one "that is firmly sustained despite what almost everyone else believes and despite what constitutes incontrovertible and obvious proof or evidence to the contrary."[23] But we need to be careful here, because the patients who have such experiences may be correctly inferring what their brains perceive.

Functional magnetic resonance imaging (fMRI) brain scans of people experiencing phantom movements[24] demonstrate that the sensory motor areas of the brain do not distinguish between imaginary and actual images and activities. There may even be a genetic encoding that creates the realistic sense of our bodies.[25] This piece of evidence allows us to look at near-death experiences, in which people who come close to clinical death maintain a realistic sense of their body, even though sensate consciousness has ceased.

To feel as though a missing limb is still there may actually have an adaptive advantage. For example, when a prosthetic device is attached to the stump of an amputated limb, people with strong phantom limb sensations adjust more quickly than those who do not experience the phantom limb. The mechanical arm or leg, as far as the brain is concerned, feels real. I once had a similar experience when I participated in a demonstration of a three-dimensional video game. In this game, one player stands on a pressure-sensitive platform, wearing a helmet and a glove that relays electronic information to and from a computer. In the helmet's visual area, the participant can "see" the other player, who in this case was portrayed as a cartoon-like figure on a three-dimensional platform floating in space. My helmet and the glove were equipped with motion sensors, so that when I turned my head, I would see different parts of the animated world; and if I looked down, I would see my animated arm holding a gun. I could wave my real arm and the simulated arm would move in the same way. This was uncanny, but it only took a few moments before my cartoon body felt real. I didn't have just a phantom limb; I had a phantom body.

Amusement parks are developing similar devices to create a realistic impression that you are hurtling through space or skiing down a mountain. Similar devices are being developed to give a surgeon's hand the sense of touching tissue deep inside a patient, even though the surgeon is in another room using a remote-control device. The military has developed similar simulations to give soldiers a realistic sense of being in battle. Some of the experiences, like losing control of your aircraft, are so real that the trainee becomes physically ill. The notion that we are "brains in a vat" is not so far from the truth.

Gorillas in Our Midst

To me, what is amazing about our perceptual process is not how accurate it is, but how real it makes the world appear. Ultimately, it is one of the most functionally useful human processes; that is why our individual sense of reality is the primary belief from which all other beliefs emerge. However, the perceptual component of belief operates in the background of our awareness. In fact, the more we consciously focus on a particular object or activity, the less we actually perceive other objects or changes in the environment. To demonstrate this phenomenon—known as "change blindness" or "selective looking"—the psychologists Daniel Simons and Christopher Chabris of Harvard asked a group of participants to watch a film of two teams of people—one wearing white outfits; the other, black—passing two orange basketballs back and forth. The participants were to count the number of times the balls were thrown by one of the teams. In the middle of the video, a person in a black gorilla suit walked into the center of the game, turned to the camera, beat its chest, and walked off.* Half of the observers did not notice the gorilla, and couldn't believe it had appeared until they were shown the film again.[26] In other experimental videos, mountains appear and disappear, colors change, and receptionists are replaced by different people, and the observer does not notice the change. Some observers even failed to notice when the heads on two people's bodies were interchanged.[27]

We make such omissions all the time. For example, you may not have noticed that the box next to this paragraph contains two unnecessary words.

> *We often do not*
> *not see what is*
> *right in front of*
> *of our face!*

Why does this occur? Mainly, it happens because it is neurologically efficient: our cognitive and organizational centers have evolved to select the least amount of information necessary to effectively respond to the world. Since there is always too much information to take in and process, the brain identifies what it believes to be significant and ignores or censors the rest. This sorting is particularly useful in emergencies, when split-second decisions are called for. Unfortunately, as many tragedies have demonstrated, the brain can overlook crucial information, particularly when we have limited

*You can view this video and other demonstrations of inattentional blindness at the website http://viscog.beckman.uiuc.edu/djs_lab/demos.html, which was created by Daniel Stevens and his colleagues at the Visual Cognition Lab at the University of Illinois.

experience in evaluating new experiences. This is why teenage drivers are so prone to accidents; their brains have not yet learned how to recognize subtle problems that can lead to disastrous situations.

Multiple Realities and Multiple Minds

The amount of information that any one brain can hold is limited by the structure and functioning of neuronal activity, but the human brain has learned to do something that no other creature has done: it permanently stores useful information outside its own neural mechanisms. We encode information in symbols that can be written on pieces of paper, bound into books, and stored in libraries. A person's body can die, but his or her knowledge can be carried by others to every corner of the Earth. And with the advent of the computer, our memory capacity becomes practically unlimited.

Perhaps the most amazing form of learning concerns the nature of the Internet, which seems to be taking on a personality of its own. According to a physics professor at the University of Notre Dame, Albert-Laszlo Barabasi, the Internet has become an extension of the human brain, functioning like a self-organizing entity. It even seems to be evolving according to the same principles that govern cells, societies, and the natural laws of the universe, without the aid of additional human intervention.[28] This, one might say, is an example of the "brain outside the vat," where the vat is a metaphor for the physical human body.

Our personal sense of reality is dictated by the sum total of all the information gathered by the human race; and as we continue to delve into the sciences, that sense of reality is bound to change. Most people find it difficult to accept the notion that the reality we perceive is not exactly the reality existing "out there," but a thorough review of the literature on perception suggests that we do not need a precise or complete representation of any object, face, or scene in order to maintain a stable impression of the world.[29] As long as nothing goes wrong with the machinery, we are fine; but the slightest trauma to the brain can interrupt neural function. Even if only a very small area is affected, a cascade of deleterious effects can ripple through the rest of the brain. This is what makes a minor stroke so potentially damaging: a microscopic death of a small number of neurons in one hemisphere can cause destructive changes in the synaptic activity of the other hemisphere.[30] When this happens, our perception of reality tilts, and our everyday beliefs collapse. But some functions can be assumed by other parts of the brain that have not been damaged, and this allows some patients to partially recover lost skills. Usually, the younger a person is, the greater the degree of recovery because the brain is still in the process of building neural connections and circuits.

For an infant, reality (as an adult knows it) doesn't even exist. Instead, chaos reigns. The infant starts out with twice as many neurons as are needed, but with far fewer neural connections. The eyes have not yet learned how to see, and the neurons that link our sensory mechanisms to other parts of the brain have yet to form the necessary connections to provide a stable perception of the world. The infant brain is like the construction site of an unfinished mansion: everywhere you turn, you'll see piles of lumber and hardware and electrical parts waiting to be slowly assembled and integrated into a useful, functional home.

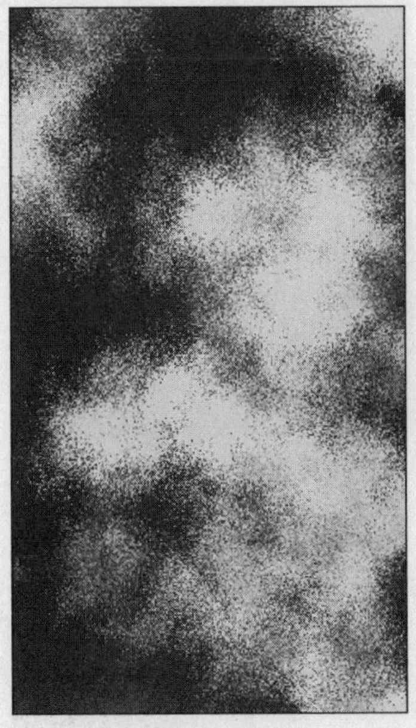

Exhibit 3-7

A newborn barely has a sense of depth perception. The sense of color is poorly developed, and the ability to focus is extraordinarily imprecise. Most of the nerve cells have not developed the myelin sheaths that are necessary for neural transmission, and the dendrites (the thousands of branches that extend from each neuron to communicate with other cells) have yet to penetrate the many layers of the brain's visual system. Exhibit 3-7 is representative of what the world looks like to a month-old infant.

By the age of three months, a baby's visual system has changed dramatically, and with the simultaneous development of memory and rudimentary language recognition, the infant can identify and recall certain objects. Shapes are still fuzzy (as depicted by Exhibit 3-8) but elements such as motion, direction, and speed can be detected and analyzed with great sophistication. This is where human imagination emerges, as the brain conjectures different possibilities of what it perceives. Is it a monster, or just a boy hugging his dog? The infant cannot be sure, and even though the parents will try to teach him or her that monsters don't exist, the possibility that such creatures may creep out of the closet will forever be imprinted in the brain. In a young child's mind—and in most adults' minds as well—fuzzy ambiguous shapes will always trigger a startle response whenever we encounter an indistinct object in the world.

Exhibit 3-8

Throughout childhood and adolescence, neuronal circuits continue to grow and die off, forming unique patterns of connections. In addition, each child grows in a different environment that can either expand or inhibit neuronal growth, which in turn can influence future personality traits. Thus a child who has a nurturing parent may perceive the world in a loving and trusting way and may treat others in the same way. But an abused child may develop an overly anxious and suspicious personality, and may be inclined to re-create abusive situations or be unable to trust others. Neurological studies have found that some abused, neglected, and traumatized children can end up with impairment in multiple brain structures and functions.[31] I do not mean to say that personality conditioning is written in stone, for even a child's brain has the potential to learn from traumatic experiences in beneficial ways.

The Two Realities of the Brain

Each hemisphere of the brain perceives reality in a different way. Generally speaking, the right side spatially grasps the wholeness of the world through feelings. The left side turns reality into sets of ideas that can be communicated through language to others. Both halves of the brain, when working together, give us a sense of reality that is clearly different from the sense formed when either side acts alone. Experiments have even found that each hemisphere creates a separate consciousness, functioning independently of the other.[32] This is why, when we are feeling calm and motivated, we might engage in an altruistic activity; but when we feel angry, we act selfishly, with little empathy or care. From a neurological perspective, each emotional state can elicit different, even opposing beliefs from one moment to the next.

Even on a microscopic level, each neuron acts independently in deciding which information, and which parts of it, to pass on. In the process of synaptic communication, each neuron changes the message slightly, excluding bits of information it considers irrelevant and adding on new bits of information. One might say that each neuron has a mind of its own, governed by its own beliefs and directives inferred by various genes. According to Bruce H. Lipton, Ph.D., a former associate professor of anatomy at the University of Wisconsin, individual cells have the molecular capacity to hold beliefs. If a cell perceives stress, it can "rewrite" the genetic directive in order to overcome the stressful condition.[33]

Imperfect Perceptions

Biological research also confirms that the human perceptual system is far from perfect. We may marvel at the complexity of the human eye, but the eyes of an octopus can see more accurately and at greater distances. A honeybee can perceive patterns on flowers that are invisible to the human eye; and many insects—which can sense polarized light—can rapidly navigate spatial realms that would be impossible for any other creature.[34] We cannot hear as well as a bat, or smell as well as a dog; but we do seem to be able to imagine different realities better than any other creature on the planet. Often, last year's science fiction becomes tomorrow's scientific fact.

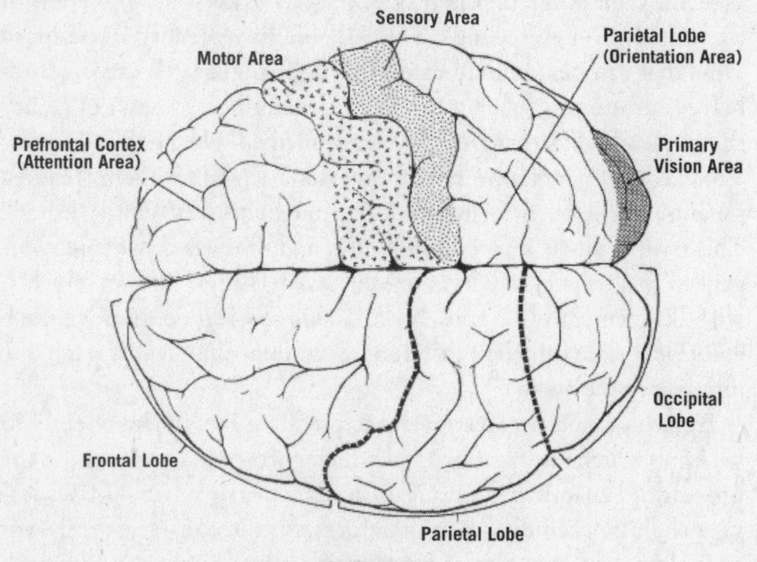

Exhibit 3-9
Top view of brain and its structures

Most of this ability relates to the frontal lobes, which are the most recently and highly evolved part of the brain, located behind your forehead. Here is where the essential mechanisms that govern consciousness reside—the repository of our most sacred beliefs and dreams. The frontal lobes (see Exhibit 3-9) have been called the "seat of the will" because they enable us to direct our attention and initiate behaviors. For that reason I sometimes refer to this part of the brain as the "attention area."* Interestingly, it is also the part that, functionally speaking, is the most removed from our direct perceptions of reality. As Rita Carter notes in her book *Exploring Consciousness:*

* In order to simplify neurological terminology, I will also refer to the parietal lobes as the orientation area and the occipital lobes as the visual processing area. However, each lobe is involved in many other perceptual and cognitive processes.

> The flow of external data, and the sensation and actions triggered by them, feed into the ordering framework of our brains and create an ever more elaborate, and ever more idiosyncratic, internal conceptual universe. It forms the likes and dislikes, habits of thought, dispositions, beliefs and memories that in time we come to think of as our "selves."[35]

This "ever more idiosyncratic internal conceptual universe" is the focus of Chapter 4, which explains how we transform perception into a meaningful set of beliefs. These beliefs, though far removed from reality, help us to deal with nearly every aspect of life. This ability highlights the tremendous power of beliefs, for existence itself, as we know it, may very well be a function that takes place only in the inner recesses of the brain.

Authors' Note:

In the following chapters, we will be discussing many complex topics concerning the neurophysiology of the brain, psychological and childhood development, morality, religious experience, mechanisms of prayer, atheism, spiritual development, and the reality-processing mechanisms of human consciousness. Each of these topics is a field unto itself, so it is impossible to do justice in one book to the wealth of research and information concerning any single issue. Our purpose is to give you, the reader, enough of an overview so that you can understand how our inner beliefs evolve. This unavoidable act of generalization, simplification, and exclusion of relevant information also reflects one of our primary conclusions about the nature of the human brain: it constantly generalizes, simplifies, and excludes relevant information in every attempt it makes to comprehend the world. This is the nature of all brains and books, which is why our grasp of reality is always incomplete.

Chapter 4

Santa Claus, Lucky Numbers, and the Magician in Our Brain: The Biology of Conceptual Beliefs

I'VE ALWAYS BEEN FOND OF SHERLOCK HOLMES'S DESCRIPTION of the cognitive architecture of the brain, which I find quite compatible with contemporary neuroscientific theory:

> *I consider that a man's brain originally is like a little empty attic, and you have to stock it with such furniture as you choose. A fool takes in all the lumber of every sort that he comes across, so that the knowledge which might be useful to him gets crowded out, or at best is jumbled up with a lot of other things so that he has a difficulty in laying his hands upon it. Now the skillful workman is very careful indeed as to what he takes into his brain-attic. He will have nothing but the tools which may help him in doing his work, but of these he has a large assortment, and all in the most perfect order. It is a mistake to think that that little room has elastic walls and can distend to any extent. Depend upon it; there comes a time when for every addition of knowledge you forget something that you knew before. It is of the highest importance, therefore, not to have useless facts elbowing out the useful ones.*[1]

Our brains do start out at birth much like an empty attic, and we spend most of our early years developing the cognitive tools that turn our perceptions into a useful and meaningful knowledge of reality. And we had better do a good job, for the older we get the less elastic our neural attic becomes.

But the brain at birth is not a proverbial "blank slate," for it comes equipped with numerous skills—like the ability to form language and perform math—which explain why most human beings form similar views of reality. On the other hand, our beliefs can be wildly divergent because what we choose to add to the neural attic will alter the way we come to understand the world.

Our beliefs depend largely on how we process the information we perceive, and these beliefs will change rapidly during the first few years of life. When my daughter was four, she once saw a magician pull a rabbit from a hat. Her eyes witnessed the event, but did she really believe that rabbits could materialize out of thin air? Was the magician controlling unseen forces, or did she suspect that he was performing some sort of trick? When I asked her what she thought, she said, with utter confidence, that some people can do those kinds of things because magic was something real. And when I asked her how magic worked, she replied—again, with equal confidence—"It just does. That's how magic works."

I wondered what she thought about other things that seemed to work mysteriously, so I asked her whether the toaster worked by magic. She laughed and said, "Of course not!" "How, then, does it operate?" I prodded. "You push the button, it turns on, and it makes toast," she replied quickly. Sensing my apparent confusion, she reassured me that magic was needed only for special things.

This is how the mind of a four-year-old works; but what my daughter did not know was that within a few years, she too would abandon most of her magical beliefs as new brain functions began to develop. It takes several decades for our cognitive abilities to mature, and during this time our belief systems, like our neuronal connections, are very flexible. But the older we get, the less flexible our beliefs become as our neural pathways stabilize. However, our ability to refine our beliefs continues to mature.

One of our earliest cognitive skills involves testing our experience

Aging, Birthdays, and Kids

It takes many years before the brain can integrate different cognitive functions in meaningful ways. At the Ben-Gurion University of the Negev in Israel, Rama Klavir and David Leiser discovered that 84 percent of children between four and seven believed that you could accelerate your age and grow up faster by having more birthdays in a year—an example of poor integration of quantitative and causal analysis. Is it possible that older people who want to ignore their birthdays are applying the same logic in reverse?

of reality. During the first two months of life, an infant does not know that an object exists when it is out of sight. Then, during the next four months of development, the notion of object permanence and continuity appears. How do scientists discern this cognitive development? They set up an experiment that allows an infant to watch a ball as it rolls behind a barrier. If you surreptitiously remove the ball, then lift the barrier, the younger infant shows no surprise. The mysterious disappearance of objects does not disturb the child. But by the age of six months, nearly all infants will show a startle response if the ball mysteriously disappears.[2] The brain, on the basis of its own biological development, expects the ball to be there.[3]

By the age of two years, a child can consciously recognize when the rules of object permanence are violated, and will usually try to figure out what went wrong by looking around the room for the ball. A four-year old might solve the perceptual mystery by attributing it to magic. Older children, however, will view the illusion as a form of deception. These responses demand increasingly complex reasoning skills that develop rapidly from the moment we are born and form the basis for certain expectations about how the world works. This is how we come to "know" the nature of reality.

Interestingly, speech recognition seems to develop earlier than visual recognition. For example, infants can recognize their mother's voice at birth, but before the age of three months they do not recognize her face.[4] And although we are born with the capacity for audiovisual integration, it takes many years of neural development for this to occur in the brain.[5]

We Tend to Believe What We Want to Believe

Our expectations have a significant influence on what we eventually believe about the world. We come to expect certain things to happen, and we expect people to behave in specific ways. These beliefs are necessary for helping us to deal with the world adaptively and productively. In our social interactions, for instance, we need to hold beliefs about what should or should not happen. For example, most children believe that if they are nice to someone, that person will respond with kindness. As we will see, these and other beliefs often develop in early childhood and are heavily dependent on our relationships with others.

It is also important that we have a good sense of what other people are thinking (this is often referred to as "theory of mind") and how their beliefs relate to ours. According to Kristine Onishi at McGill University and Renée Baillargeon at the University of Illinois, children as young as fifteen months can begin to recognize other people's beliefs—including false ones—about various situations and use that information to interpret how others will respond.[6]

However, neither children nor adults have a well-developed capacity to distinguish the accuracy of their own beliefs. In fact, adults are particularly vulnerable with regard to maintaining self-deceptive beliefs, especially when comparing their own intelligence and attractiveness with other people's.[7] For example, in various surveys conducted over the years, approximately 90 percent of the respondents believed that they were smarter, healthier, and more industrious than the average individual. My favorite example involved a survey of university professors: 94 percent believed that they were better at their jobs than their colleagues. Statistically, nearly half of those professors would have to be wrong. Most people, in fact, overestimate their personal abilities, and unfortunately their inflated beliefs cause them to suspend their ability to test reality.[8] Smokers underestimate their risk of lung cancer.[9] Business managers make overly optimistic forecasts, leading their organizations into initiatives that usually fail or fall short of expectations.[10]

But pessimistic beliefs, even though they may be more realistic, are stressful, and too much stress releases a cascade of destructive hormones that can seriously compromise a person's health.[11] Furthermore, too much pessimism can lead to depression, which suppresses the functioning of essential neurotransmitters. This, in turn, leads to physical inactivity, instability of moods, and a number of physical symptoms and diseases. Thus it seems that the brain, in its innate wisdom, biases us toward optimistic beliefs. If Edison, Beethoven, and Michelangelo had given up in the face of adversity, the world would be a poorer place.

This illustration shows how fourteenth-century physicians envisioned different cognitive functions of the brain, a concept dating back to Hippocrates (460–377 B.C.). He argued that all our "pleasures, joys, laughter and jests, as well as our sorrows, pains, grieves, and tears" arise from the brain. "It makes us mad or delirious, inspires us with dread and fear, whether by night or by day, brings sleeplessness, inopportune mistakes, aimless anxieties, absentmindedness, and acts that are contrary to habit."

Optimism can be very beneficial, helping us to overcome situations that seem difficult or threatening; and extreme optimism concerning recovery from a life-threatening disease may make the difference between survival and death, since positive beliefs can stimulate the immune system in healthy ways. In one study, conducted by psychologists at the University of California, Los Angeles, ninety first-year law students were evaluated to see how optimistic beliefs affected key immune cells and mood. Those who felt confident about their abilities and expectations of success had more helper T cells, which support immune responses; and more effective natural killer cells, which destroy substances that are poisonous to cells.[12] In other studies, optimistic people have been found to secrete less cortisol, a stress-related hormone that suppresses the immune system's function.[13] Lower cortisol levels make the immune system more effective.

The Emergence of Cognitive Beliefs

Once we believe that our perceptions accurately represent something in reality, the brain begins to send this information through a hierarchical processing system that allows us to compare the representation with our memories and other beliefs. These cognitive functions, which are largely preconscious, are the real magicians of the brain.

Eugene D'Aquili and I postulated and found evidence for specific cognitive processes that are not only essential to the formulation of everyday beliefs but also responsible for the emergence of spiritual perceptions, mystical experiences, and unitive states of consciousness.[14] These cognitive processes include (1) the abstractive function, (2) the quantitative function, (3) the cause-and-effect function, (4) the dualistic or oppositional function, (5) the reductionist function, and (6) the holistic function, which, in essence, puts everything together into a meaningful, comprehensive worldview. Let us look at how each of these functions works to help us form our everyday beliefs.

Labeling the Universe

Although we are not aware of it, our brain spends a great deal of time labeling everything we perceive. As parents, we invent games for our children to ease the burden of this essential learning process. For example, my daughter and I play "I spy" whenever we're walking in the neighborhood or driving in the car. I'll start by picking out a nearby object, but I won't say what it is. Instead, I'll announce, in my best Sherlock-Holmesian way, "I spy something green and brown." If my daughter guesses wrong, I'll add another clue—"I spy something with branches"—until she identifies what I see and proudly proclaims, "A tree!"

On a conscious level, I'm helping my daughter to associate a series of abstract concepts with specific objects in the world. Our brain does something similar with each perception: it converts the neural stimulation into various categories that represent lines, depth, color, shape, texture, and so on, until it puts together an abstract concept that we will later call a tree. All this happens without language or words, for at this stage a tree is nothing more than a distinctive pattern of neuronal flashes and blips. This abstractive function, also known as object representation, is responsible for giving us the sense of "thingness" that allows us to distinguish one object from another. But this distinction is an arbitrary invention, because our actual perception of the world is experienced as undefined data coming in through our senses. The information doesn't become a "dog" or a "cat" until the abstract function of the brain goes to work.

It takes a child months of intense learning to identify anything at all. The brain must integrate its perception of movement, sound, and tactile experience before it can even identify an object as a potential animal, plant, or rock. Then the brain must construct dozens of other categories and labels such as hair, legs, eyes, barking, and a wagging tail before the nonverbal concept of "dog" is formed. Finally, the brain must send this information to other centers where it will be turned into the auditory, written, and pronounceable word "dog." Snakes have it easier. When they perceive movement, it means eat, strike, or slither away. Their tiny brain cannot form categories such as "friend"; thus pet reptiles cannot grasp the notion that a human caretaker is feeding or protecting them. But dogs can, demonstrating that they conceive of a wider range of categorical skills. For example, my dog believes that if he barks while standing near his bowl, I will come and fill it with food. But he can't understand why, when I'm watching a football game, I wait until a commercial to fill his urgent request. "Food," "bowl," and "hungry now" are categories of conceptualization that canines grasp, but other concepts, like "waiting" or "sports addict" are incomprehensible, at least to my dog.

The brain loves to label everything, so much so that by the time we reach the age of six, we hate to encounter objects that are unidentifiable. In second grade, as I recall, my friends and I used to put a handful of soggy macaroni in a brown paper bag and ask some unsuspecting first-grader to reach inside and touch it. Few were willing to do so, but for those who could not turn down a dare, we'd inform them, just as they made contact, that it was a pile of human brains. Out flew the hand immediately. (Actually, only the girls withdrew their hands; the boys could not have cared less.) The same thing happens when you put something in your mouth that you can't immediately identify—your body instinctively wants to spit it out. After all, our sense of taste developed to help us to survive.

Abstraction acts as a doorway between direct perception and consciousness, for we depend on our concepts, labels, and words to shape our awareness. This is problematic when it comes to spiritual matters, which, by definition, refer to realms that have no physical reality. How would you play "I spy" when it comes to a concept such as God? What categories would you use, since shape, size, color, and location would not apply? We usually end up giving our

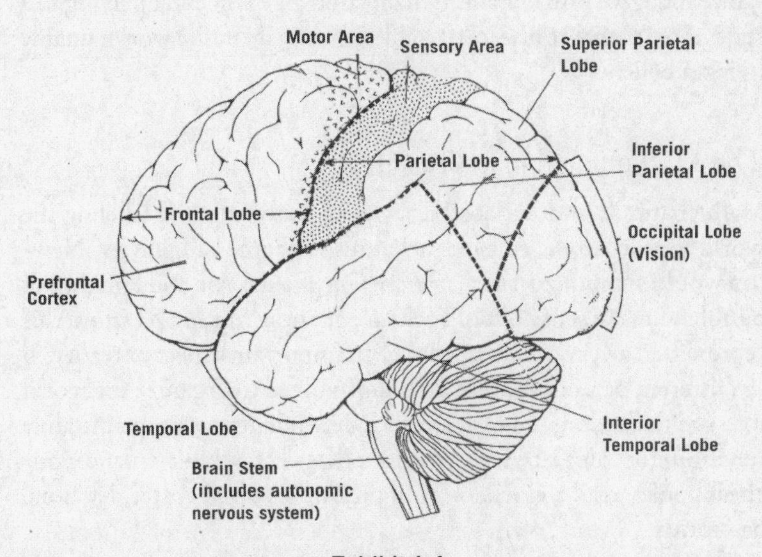

Exhibit 4-1
Side view of the brain and its structures

children images they can relate to, like Michelangelo's painting on the ceiling of the Sistine Chapel. Young children can form categories for concrete objects, but they have enormous difficulties with abstract concepts like freedom, fairness, or God.

The neural structure related to abstraction is probably located at the junction of the superior temporal lobe and the inferior parietal lobe in the left hemisphere (see Exhibit 4-1). This area can generate classifications of objects that are vastly more complex than those in other sensory areas. Patients with injuries in this area may not be able to identify objects placed in front of them and may have difficulty writing, comprehending language, or performing mathematical computations, all of which depend on the labeling of objects and abstract concepts. Alzheimer's disease similarly affects abstract classification processes, making it difficult for patients to express what they are thinking or what they want.

Our brain, then, transforms reality into abstract categories and labels, and these labels are, by definition, nothing more than intangible beliefs, assumptions about a world that cannot be directly perceived. In this sense, labels, beliefs, and reality are one and the same; and if we lose our ability to abstract, we will end up living in a state of perpetual confusion, unable to navigate in the world, unable to form beliefs.

The Mathematician in Our Head

Mathematics is another abstract way of analyzing and labeling the world. For example, consider what most people call gravity. Newton would symbolize it as $F_{12} = F_{21} = Gm_1m_2/R^2$. You and I might not be able to make sense out of such an equation, but for Newton, this represented a very accurate view of an important aspect of reality.

Different parts of the human brain evolved to organize the world into such equations, and we are very dependent on this quantitative function for our survival. Without it, we couldn't balance our checkbooks, read a clock, or even pick up a cup of coffee, let alone pay for it.

The brain encodes certain information into units of quantity, and, in conjunction with our language centers, assigns numeric labels to give us the concepts 1, 2, 3, etc. Ongoing research continues to demonstrate that we are born with the capacity to quantify objects and concepts;[15] that a single, abstract system of number representation is present and functional in infancy;[16] and that we may be genetically endowed with the ability to think mathematically.[17] Even during the first weeks of life, infants appear to have the ability to discriminate between different groups of objects.[18] By the age of six months, infants can distinguish between large numbers of objects in a way that demonstrates an innate comprehension of ratios (i.e., the relationship between two quantities of objects).[19] From nine to eleven months, infants can even perform the rudimentary functions of addition and subtraction.[20]

This ability to assign and sequence numeric values is not limited to human beings. Even an ant can apply mathematical equations to the task of building nests[21] and searching for food.[22] Although some ants create and follow pheromone scent trails, others use computational skills to map out their locations[23] and find quicker, more direct ways to travel to and from their nests. After an ant finds your picnic basket, it can follow the algorithms it constructed in its minuscule brain and make its way directly back to its nest, something that very few taxi drivers can do.

Humans also assign different emotional values to numbers. We rank hotels and restaurants, giving them, say, one to five stars for service and quality. We grade everything, from our kids in school to our Internet connection speed, and we get very upset when our numerical expectations are not met, especially when these expectations concern how much money we have. We even consider certain numbers more lucky and powerful than others. In other words, we imbue the "lucky" numbers with magical properties. Western cultures associate the number thirteen with bad luck; this is why you'll rarely find a thirteenth floor in an American hotel. In China, however, the symbol that denotes thirteen also means "must be alive" and is therefore lucky. But you might not find a fourth floor in Chinese high-rises, because four implies death. For similar reasons, the Japanese do not like the number nine: it has the same pronunciation

John Dee (1527–1608) was a man obsessed by the number seven. He was a royal astrologer to Queen Elizabeth, and his fascination with numerology led him to create the mystical symbol known as the Sigillum Dei Aemeth, shown here, which contained the seven names of God. Not only was he a brilliant mathematician who introduced Euclidean geometry to the English-speaking world; he was also an excellent spy, signing his espionage notes 007. He was Ian Fleming's prototype for James Bond, and his long white beard and love for alchemy provided the inspiration for dozens of future literary wizards, including Tolkien's Gandalf and Rowling's Dumbledore. Dee may have modeled his own image on the Arthurian legend of Merlin.

as "ku," a word connoting agony and torture. Seven may be lucky in Las Vegas, but in Chad, Nigeria, and Benin, all odd numbers are bad.

Such beliefs, like other superstitions, are also grounded in myths passed down by previous generations; and the more frequently they are repeated, the more difficult they are to ignore. Why human beings assign positive and negative attributes to certain numbers is neurologically unknown, but it is most likely related to the brain's propensity to assign emotional values to everything. When an important event is associated with a quantified value, that number takes on increased significance and meaning, even if the association is coincidental. I recall quite vividly when I was a child in school, being the third person called on to spell a difficult word in a spelling bee. I succeeded and won. In that same year, my baseball coach had me bat third in the lineup, and I hit three home runs. My soccer coach gave me a jersey with the number 3 and our team won the championship that year. Three became my lucky number, and even today, when I encounter it, I have the uncanny feeling that something good is about to happen.

Quantifying the world is so important to the functioning of our brains that it becomes integrated into every aspect of our lives. It influences our architecture and art, and it fills our religious rituals with significance. Hindus pray three times a day to the Vedic gods; Muslims pray five times a day; Roman Catholics are directed to pray seven times a day; and an orthodox Jew is expected to give praise to God 100 times a day. Indeed, many religious texts, including the Old Testament, have passages that are filled with quantitative and numerological beliefs.

Numbers impart meaning. That is why people pay attention to statistics, stock-market reports, and death tolls. And the larger the numbers, the more impressive the associated information seems. We are awed by the immense age of the universe, or Bill Gates's wealth, or the number of neurons in the brain—100 billion, which coincidently matches the number of stars in our galaxy. As for the number of synaptic connections in the human brain, the estimates fall somewhere between 100 trillion and 1 quadrillion. One quadrillion is written as a 1 followed by 15 zeros:

1,000,000,000,000,000

Even more impressive is a googol, which is greater than the number of particles in the known universe:

10,000,000,000,000,000,000,000,000,000,000,000,000,000,000,
000,000,000,000,000,000,000,000,000,000,000,000,000,
000,000,000,000,000,000,000

All numbers seem magical because they represent a complex world in simple figures. But statistics never impart absolute truths about the world.[24] They merely compare objects or other abstract symbols, and they can be easily manipulated in ways that distort reality. We'll talk about this more in Chapter 10, but for now, since you'll be encountering many statistics throughout this book, it's wise to be cautious when you feel the influential pull of numbers.

Quantitative functions are processed mostly in the inferior parietal lobe, adjacent to the areas that support the abstract functions mentioned earlier. When things go wrong with this part of the brain, you may lose the ability to move safely in the environment. You can't calculate the distance between the door and your bed, or clearly recognize the position of your arms and legs. But if you deliberately decrease activity in this area through intensive meditation or prayer, you can decrease your quantitative sense of time and space and thus create an experience that mirrors the timeless mysti-

Causal Organs, Ancient and Modern

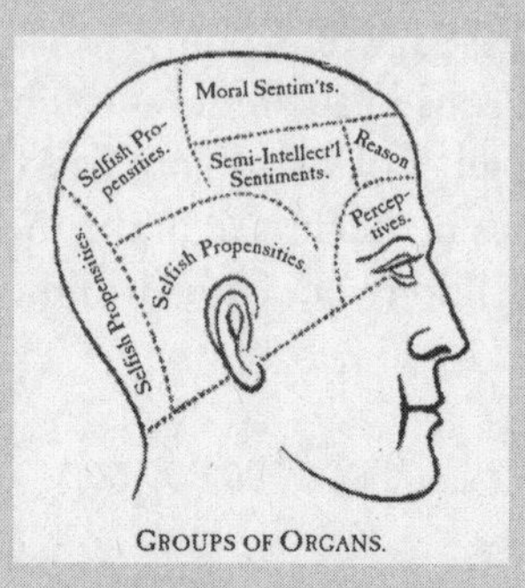

GROUPS OF ORGANS.

George Combe, in his book *A System of Phrenology* (1853), posited the notion of a "causal organ" in the brain "lying immediately at the sides of the organ of Comparison" that allows us to perceive the sequences of event. The picture here is from *New Physiognomy* (1871) by Samuel Wells. Cause-and-effect processes have also been attributed to various genetic properties, but the biologist Andreas Wagner, at the University of New Mexico, believes that many genetic processes are nonlinear; thus no notion of causality is likely to apply to them. (See "Causality in Complex Systems," *Biology and Philosophy*, vol. 14, 1999.)

cal states of consciousness so often described in spiritual and religious texts.

If There Is an Effect, There Must Have Been a Cause

Another essential function of the brain is its power to organize neural information into a sequence of causes and effects. Without this process of "neural logic,"[25] the events of the day would seem chaotic and meaningless. Much of the logic of the brain occurs unconsciously, but we can see it operate consciously whenever we ask "Why?" In fact, most of our expectations, predictions, and beliefs depend on our ability to build cause-and-effect associations between thoughts and events in the world. Language itself must follow a sophisticated series of cause-and-effect rules, for if it didn't, your words could not be understood by others.

Patients who have had a stroke or who have tumors in the areas associated with causal functioning often have difficulty determining why something happened. They feel lost, even terrified, and they have difficulty making decisions because their ability to make logical predictions about the future has been impaired. And if their language centers are affected, their speech may be incomprehensible.

Cause-and-effect associations, in conjunction with other cognitive and executive functions, help us to order our behavior and thoughts while simultaneously decreasing our anxiety and fears.[26] But when our ability to form causal associations is disrupted, whether through illness, drugs, trauma, or a misperception of reality, the emotional centers of the brain begin to fire rapidly, alerting us that something is amiss.

Let's take a look at how this causal function might operate on an average morning. The alarm clock goes off, and the sleeping brain is shocked into consciousness. If you were sound asleep, your limbic system might make you throw the damn clock across the room; but if your frontal lobes come online and remind you that you do need to get up, you'll probably just reach over and turn it off. Slowly, a series of "if-then" beliefs or scenarios will drift into consciousness: "If I slept an extra half-hour, I'd feel good, but then I'd probably be late for work. The boss said that if I was late again, I'd be fired. If I'm fired, then I won't be able to pay my rent." At this point, your frontal lobes will probably persuade you to climb out of bed.

Now let's see what happens if something goes wrong in your brain and your causal functions don't turn on. You'll hear the clock buzzing, but you won't figure out what it is. Worse, you probably won't know where you are. All this will drive you nuts, so you'll throw off your suit of armor and fly out the window, only to find yourself surrounded by a horde of flesh-eating zombies. Living in such an illogical world would be a nightmare. Fortunately, when we have these perceptions, we usually wake up and realize that they were just a bad dream. But some people have damaged causal functions and cannot tell if they are dreaming or awake.

Something similar happens during sleep paralysis, a disturbing phenomenon that disrupts causal reasoning. The patient wakes up but cannot move and continues to have various visions and bodily experiences.[27] Some people feel the presence of an unusual entity or intruder, and are pulled into terrifying realms filled with auditory and visual hallucinations. Others have reported a demonic presence sitting on or in their chest, causing them to feel as if they are being suffocated. A few, however, experience floating and flying sensations, out-of-body experiences, ecstasy, or bliss. These individuals may feel deeply transformed by the event, which alters their spiritual beliefs.

Some scholars suggest that historically, the experiences encountered during sleep paralysis may account for a variety of descriptions of saintly and demonic possession, as well as reports by people who believe they were abducted by aliens. Neurologically, there is a loss of sensory information reaching the brain. The parts of the brain that generate consciousness react to these ambiguous perceptions of touch, sight, and sound by constructing extraordinary scenarios that attempt to make sense of the unusual experience.[28] Sleep paralysis is not rare. Some studies suggest that it occurs in 15 to 20 percent of the population, regardless of sex, culture, or age.

The brain-scan studies that I am now conducting at the University of Pennsylvania have shown that religious practitioners and meditators have altered neural activity in the frontal and parietal lobes (and in other areas associated with language and emotion) that is associated with a temporary suspension of their awareness of cause and effect. In this altered state of consciousness, everything exists "in the moment." Reality takes on an otherworldly quality, and many practitioners describe a feeling of being overtaken by an immediate "presence." These altered states are usually temporary, and when the practitioner returns to his or her everyday awareness, causal functioning resumes and the brain attempts to integrate the experience with the practitioner's primary beliefs. Thus, a Buddhist meditator would interpret the experience in accordance with Buddhist ideas (feeling connected with absolute consciousness, for example), whereas a Franciscan nun would interpret her experience through the Catholic lens (as being in the presence of God).

When weird things happen, the causal functions of our brain try to make sense out of them—to give them a reason, a cause. If we cannot figure something out, we might call the event magic, or a miracle, or a hallucination, depending on which explanation makes more sense or brings us more peace or hope. Usually, however, when we can't make a causal connection, we feel intensely disconnected and isolated from the world. Neural disorders can also give us this feeling. For instance, they can cause us to believe that we have done one thing when we've done another. One patient I knew continually burned herself because she could not connect the idea of "stove" with the experience of "hot." Sometimes patients who can't see cause-and-effect relationships will go to extremes, confabulating stories to explain discrepant events. Patients with Korsakoff's syndrome, for instance, might imagine that they spent the weekend in a foreign city, when, in fact, they were bedridden in a hospital. In such patients' consciousness, there is a glitch in the memory processes of the frontal lobe, which interrupts the ability to distinguish between imaginary and experienced events.[29] The patients fill in the blanks and end up believing that their fantasies are real.

Like other cognitive skills, causal reasoning begins in infancy and develops gradually as new experiences accumulate. Children younger than three must depend on adults to help them build if-then connections as they learn to explore the world. When a toddler turns a glass of milk upside down, giggling as the milk splashes across the floor, he or she is simply working out the laws of gravity, as well as the laws of mom: "Things fall, if they're not held up, and mothers get mad over spilled milk." Voilà! Two more causal beliefs are confirmed.

If-then scenarios are also responsible for many erroneous beliefs. If you wake up with a cold, you might take an herbal remedy or a vitamin pill; and if you feel better the next day, you will probably believe that your remedy worked. If you are given hard evidence that your remedy does not work, you will probably ignore it. On the other hand, if you don't get well, or if you are worried about your child's cold, you might insist that your doctor prescribe an antibiotic. The doctor may know that you have a virus, and that antibiotics will work only against bacterial infections, but most doctors go along with such demands. One study found that 42 percent of the physician respondents inappropriately prescribed antibiotics for the common cold.[30]

Pascal's Wager

Blaise Pascal (1623–1662), in trying to decide whether to believe or disbelieve in God, used causal reasoning. His argument is known as Pascal's wager. He concluded that it was wiser to maintain a belief in God. Briefly, if you believe in God and there is a God, you are OK. If you believe in God and there is no God, you are also OK. If you don't believe in God and there is no God, you are still OK. But if you don't believe in God and there is a God, then you might be in trouble, facing an eternity in hell. Logically, the odds are in your favor if you believe. Of course, if the god you believe in is vindictive, you could find yourself consumed by fear or driven to destructive acts.

People will usually recover from many minor illnesses in a couple of days whether they take anything or not; but once they have made a causal association and have committed it to memory, they become convinced that the remedy they took brought about the cure. James Alcock of York University in Toronto calls this "superstitious conditioning."[31]

Trusting other people's opinions is also based on causal reasoning: "If everybody believes such-and-such, then such-and-such must be true." After all, if 95 percent of Americans believe in a spiritual reality, they couldn't all be wrong, could they? Science, too, is primarily based on causal beliefs, for if research does not make sense in some logical and reasonable way, other researchers will consider the conclusions false, or at best without merit, even if the results are actually true.

Drawing causal connections helps us to predict the future, but it also impels us to speculate about the past—to explore our biological roots, as well as the earliest causes of our existence. The "cosmological argument," or "first-cause argument," was embraced by Aristotle, who believed that everything that happens is caused by something else. In western religion, "something else" is God. For astronomers, the big bang theory accounts for the beginning of the universe. The causal nature of the human mind makes some astronomers and theologians ask yet another question: "What caused the big bang, and what existed before it?" It seems as though the human brain never stops asking causal questions.

Us versus Them: Dividing the World into Opposing Dyads

In the process of organizing, labeling, and quantifying the world, the brain has a tendency to reduce everything to as few components as possible. In the hidden recesses of the inferior parietal lobe, there exists a cognitive function that puts abstract concepts into polarized dyads, or dualistic terms. Think back to your early school days, when many of the vocabulary words were presented as opposites, such as "up versus down," "in versus out," and "on versus off." Many scientists believe that neural processes themselves are dualistic, turning on and off in much the same way that a computer program uses a two-digit code (where 0 = off and 1 = on) to analyze information. For example, within the visual system, there are neurons that fire only when an object is moving in one direction, while others fire only when the object moves in the opposite direction. Pain and memory also have been linked to binary on-off mechanisms within the brain.[32]

In terms of neurological development, it is easier for the brain to first quantify objects into pairs, and then to differentiate them into opposing groups: light or dark, happy or sad, fact or fiction, good or evil, right or wrong, Republican or Democrat, and so on. Together, such dyads represent a unified concept. For example, you cannot conceive of "up" without making reference to "down," because each term is defined according to its relationship to the other.

In the Beginning, God Created Opposites

The Old Testament is filled with examples of oppositional beliefs. In the first few pages of Genesis, God creates heaven and earth, separating the light from the dark and the land from the water. Then God creates man and woman, who discover the nature of good and evil by eating from the tree of knowledge. The neural processes that create opposites are often expressed in religious literature and theology.

The brain tends to reduce cause-and-effect cognition into dualistic if-then scenarios because these are an easy, neurologically efficient way to make sense of the world. When faced with a wild animal, we will first freeze to evaluate the situation (is it dangerous or safe?), and then we will react (the fight-or-flight response). If we had a particular love for the creature, we might attempt to befriend it, but that requires far more neural activity. And in the case of a poisonous reptile, it might not be the choice most likely to ensure our survival.

This neural process of simplification and generalization is, in effect, a form of biological stereotyping because it does not take into account individual differences and nuances. Furthermore, once an oppositional dyad is created, the brain will then impose an emotional bias on each part of the dyad. Thus, once we divide objects, people, and ideals into groups, we will tend to express a preference for one and a dislike for the other. We will root for our favorite baseball team and disparage the challenging team, and we will tend to distrust whatever the opposing political party says. Unfortunately, such biases are not limited to politics and sports; we also assign preferences and dislikes to people from different cultural, religious, and ethnic backgrounds. As hundreds of studies have confirmed, the in-group will always develop scenarios—pass laws, distribute benefits, etc.—that are less than favorable for the out-group.[33]

This "us-versus-them" mentality exists even when the division is arbitrarily assigned. Systematic research by Henri Tajfel, which began in the 1970s, found that when individuals are randomly placed into different groups, they feel stronger about their own group and tend to feel negatively about other groups, even when issues of religion, sexual identity, and culture are factored out.[34] Thus simply being a part of a group results in ill will toward other groups. The human origins for such beliefs most likely evolved from the defensive and aggressive behaviors that different animals use to compete for control over territory, food, and mates. We are biologically prone to divide people into groups, to categorize and stereotype them, and then to evaluate them in preferential and prejudicial ways.

This inborn "us-versus-them" mentality can be easily converted into racism, as was dramatically demonstrated in a well-known experiment conducted by an elementary school teacher, Jane Elliott, after the assassination of Martin Luther King, Jr. One day, she told her all-white midwestern class of students that they were going to divide into two groups: blue-eyed and brown-eyed children. On that first day, she told them that blue-eyed children were smarter, nicer, and neater than those with brown eyes. She praised the blue-eyed children and gave them special privileges. Within hours, the blue-eyed children began to torment those with brown eyes, even former friends. The next day, the teacher told the class that, actually, the brown-eyed children were better than the blue-eyed children. On both days, academic performance declined in the group that was being told it was inferior and was being discriminated against. Afterward, the children discussed the effects of discrimination and how they felt. Decades later, some townspeople still were angry about the experiment, but many of the former pupils, now grown up, remembered the profound lesson they had learned and credited Elliott with helping them understand the pain of discrimination and the importance of compassion.

Brain-scan studies have found that the amygdala—the neural part of our emotional system that registers fear—does react when we first observe a person from a different ethnic background; but a closer examination of the most recent studies shows that there is significant variance in this response, and our responses are based on a variety of internal and external influences.[35] In one study, the brain's initial reaction decreased in less than half a second,[36] suggesting that we are innately predisposed to temper hostile reactions. In other words, the brain also has built-in mechanisms for suppressing, ignoring, or overriding its initial startle response. Racism, however, is a hostile form of discrimination that equates biological and genetic tendencies with certain values.

We can teach our children not to reject others automatically because of race, sex, wardrobe, or economic circumstances. One technique that is used in schools to interrupt oppositional beliefs is called the jigsaw classroom experience: when children are placed in groups with other minorities, and given a project that requires everyone's assistance, prejudices fall away, hostility fades, and group cooperation flourishes.[37]

When we are faced with any belief that conflicts with our own, it takes additional effort and time to override these biologically based cognitive biases, but by doing so, we can become more open-minded. We can even reach a point where we realize that many notions such as good and evil are largely arbitrary and relative, depending on many conditions and circumstances. History, of course, reminds us how hard it is to maintain an open mind, especially in times of war, when, from Troy to Gettysburg to Iraq, the position of both sides is the same: "We are assured of the rightness of our cause, and confident of the victories to come."[38]

Seeing the Tree within the Forest

The reductionist function of the brain allows us to take our notion of reality—or any concept, for that matter—and break it down into distinct, separate components. In a sense, this function is like a powerful magnifying glass. If you look at a speck of dust under the lens of a microscope, a whole new world is revealed. The closer you look, the more details you will find. We need to rely on reductionist processes in order to analyze or quantify anything. Without them, we could not develop cause-and-effect scenarios, or understand anything about our world. There's an old saying about not being able to see the forest for the trees; but without reductionism you wouldn't be able to see a tree within the forest—let alone its leaves, branches, or roots.

Scientific research depends on reductionist principles: such research takes something large, like the universe, and breaks it down into galaxies, solar systems, planets, minerals, molecules, atoms, and quarks. Newton did this when he envisioned the universe as a finely constructed, rationally ordered machine, created by God and governed by simple mathematical laws. If we reduce the whole to the parts, the world becomes more comprehensible and manageable. However, reductionist frameworks, and the beliefs they generate, can give us only a partial view of reality. Einstein realized this when he pointed out the fallacies of a purely mechanistic and deterministic worldview. In his relativistic universe, clocks can run slower or faster, and distances can be stretched or shrunk. Fortunately, most of our lives are spent in a Newtonian world where time and space do not bend.

If reductionist thinking is taken to extremes, you can become so absorbed in the details that you forget about the larger world. In other words, you won't see the forest *because* of the trees. This happens when we lose ourselves in the details of a problem at our work; we can miss the "big picture" and fail to find a solution that works. It's like looking at an impressionistic or pointillist painting of a daisy; when you get too close, you see the many brushstrokes and dots, but not the flower.

Obsessive-compulsive disorders (OCDs) reflect an overly reductionist brain. OCD patients become lost in a labyrinth of details, and in order to control the anxiety that this situation generates, they resort to complex rituals that seem unusual or meaningless to others but are designed to organize and control chaotic feelings and thoughts. OCD individuals often develop rigid systems of beliefs, which essentially act as a defense mechanism to prevent them from feeling overwhelmed.

Religiously preoccupied individuals follow a similar cognitive strategy, but their beliefs become limited to the object of their obsession. Such people might spend years morbidly ruminating on a few moral scruples and lose the overall context of their beliefs. Although they usually understand every detail of their religious doctrine, they never achieve a sense of satisfaction or accomplishment. Instead, they constantly pray, or ritually punish themselves.

The reductionist function most likely occurs in several different areas of the brain, including the inferior parietal lobe, the superior temporal lobe, and parts of the frontal lobe. In addition, the left side of the brain seems to be more active in providing us with our reductionist analysis of the world.[39]

Seeing the Forest in the Trees

In contrast to the reductionist function, which breaks the whole into its parts, our brain also comes equipped with a holistic function, which takes the sum of the parts and perceives them as a whole. When we look at the world holistically, all things appear to be deeply interconnected. This capacity, by the way, is not unique to human beings.[40] In fact, holistic perception may be the basic form of sensory organization for other living organisms. However, the higher cognitive functions we have described above seem to be very limited in nonhuman species. For example, human listeners process auditory sequences holistically, whereas nonhuman listeners focus on specific aspects of individual tones.[41]

Current research has identified specific neurons that are sensitive to holistic representations; however, the majority of our neurons reduce and categorize experience into object fragments.[42] This may explain why reductionist thinking is so predominant in human culture, and why holistic perceptions are so often dismissed skeptically.

Logical, rational, and reductionist processes are primarily carried out by the left side of the brain, whereas the ability to view the world in an integrated and connected way is associated with the right side of the brain. Because holistic processes are organized differently, they provide us with a very different view of the world. They are not reducible into parts, nor are they associated with causal awareness.

Instead, they feel intuitively correct. We see the forest instead of the trees. We grasp the big picture, but not the details. For example, if we see a group of dark-skinned people who are speaking a language we don't recognize and dancing around a fire, we might make the holistic assumption that they are a tribal people with shamanic beliefs. Of course, we'd have to use reductionist processes to identify the specific tribe, and when we do so, we might discover that our holistic conclusion was wrong. In the example just given, it might turn out that we were observing Mongolian shamans performing an exorcism, or a group of Tibetan priests engaged in a holiday celebration.

On an unconscious level, processes such as facial recognition rely heavily on the holistic functions of the brain.[43] This explains why it is so difficult to describe a person's facial features accurately, even if that person is a family member or a close friend. We may have a global sense of what the person looks like, but we do not have the precise words to articulate it.

Spiritual experiences also seem to rely on the holistic functions of the brain; this may explain why individuals describe such experiences in broad, sweeping, poorly defined terms. In fact, words like "enlightenment," "transcendence," and "spirituality" have proved very difficult for scholars to define because holistic processes do not operate according to the principles that govern the other cognitive functions of the brain. When holistic processing predominates, we consciously do not feel a very strong need to analyze, compare, quantify, or justify our perceptions or beliefs, because everything feels "just right."

Doctors, too, depend on holistic functions every time they make a diagnosis. For example, when a person is admitted to our hospital, I will look at the patient's charts and test results as I search for a possible tumor, a stroke, or some other abnormality to explain the symptoms. Often, my evaluation will be based on an overall impression about what is happening in the patient's body, but I can't identify a specific feature or fact that led me to my conclusion. Because holistic functions are not language-based, they tend to be more difficult to communicate or define. You can imagine how difficult it is to teach medical interns how to develop such intuitive skills.

Buddhist meditation attempts to teach practitioners how to accentuate holistic awareness by temporarily suspending the processes of logic. In comparison, western religious practices tend to be more reductionist, often applying causal reasoning and logic to build a theology. However, western religious traditions also have rich mystical elements and a holistic awareness in which the practitioner can feel connected with the universe or God.

Does holistic thinking offer a more accurate or integrated view of the world? A few years ago, our radiology staff had a chance to test this hypothesis in a very particular way. We compared the differences between human and computer evaluations of brain scans, thinking that a computer might more accurately quantify which parts of the brain were not functioning normally, leading to a more precise diagnosis. To our surprise, humans, using their intuitive holistic skills, did slightly better than the computer, most likely because we derive information from various patterns in the entire scan that help us in our diagnosis.

Thus we are born with two hemispheres that will, over time, develop two distinctly different worldviews, and our consciousness does not seem to like this very much. The right side of the brain wants to exclaim, "This is it—this is the whole picture." But the left side intrudes, saying, "No, you're wrong—look at all the inconsistencies and differences and pieces of the puzzle that have been ignored." Neurological studies seem to confirm this inner conflict by showing that each person is capable of thinking both ways, though not necessarily at the same time.

However, the brain also depends on maintaining two distinct worldviews, because people who have suffered from a stroke to either side of the brain often find themselves living in a disjointed world. For some, that world will be logical and reasonable, but with no sense of meaning or purpose; others may find themselves feeling peaceful and contented, but lacking any intellectual interest or motivation. Thus, we need both sides of our brain and both functions to feel satisfied and whole.

Holistic awareness, like other forms of subjective experience, may be an emergent property that arises from a vast network of interconnecting neural processes. Self-recognition of this internal experience might even be the ultimate expression of the holistic function, when we realize that the parts (the neurons, the cognitive processes, etc.) can be considered only in relationship to the whole. This, then, would be what gives us an overall feeling of satisfaction and of being part of the world.

Santa Claus, Little People, and the Easter Bunny

These six cognitive functions—abstractive, quantitative, cause-and-effect, dualistic-oppositional, reductionist, and holistic—work in conjunction with many other neural processes to create our belief systems. Some beliefs will have strong emotional value and thus will be deeply embedded in our memories; other beliefs will elicit only a mild response or reaction. If a concept or experience elicits no emotional response, it probably will not reach the level of consciousness.

Let's take a moment to illustrate how beliefs are shaped as they pass through the various processing centers of the brain. Why, for example, do most American adults believe in God but not in the existence of a jolly old man who sneaks down chimneys on Christmas eve? Young children usually believe in both—along with an array of fairies, goblins, and elves—but as they grow older, they come to the conclusion (usually by the age of four or five) that Santa and the little people are nothing more than myths.[44] How does this transformation occur?

Do You Believe in Little People?

On October 28, 2004, *Nature* announced the discovery of a lost race of humans, Hobbit-size creatures (Homo floresiensis) that were only three feet tall. In response, Christopher Chippindale, curator of the University of Cambridge Museum of Archaeology and Anthropology, commented, "Is there truth after all in the many stories from many lands of other humans, extralarge or extrasmall, living in the mountains or the forests, which have been dismissed as myths and fantasies?"

First, most Christian parents introduce their children to the idea of Santa Claus, which is reinforced through cartoons, movies, and storybooks. Many children—and many adults, for that matter—have been captivated by the Grinch who stole Christmas. Such tales are filled with emotional excitement because they often pit evil against good, helping the child to develop oppositional beliefs. With Santa, there's an even stronger emotional incentive to believe: seeing all those presents under the tree. The combination of pleasure, mystery, excitement, and anticipation becomes fused with the image of Santa Claus and is permanently embedded in the memory circuits of the brain. Then, various causal connections are made. For example, a child might think, "If I believe in Santa, and leave him cookies and milk, I'll get all the presents I want."

By the age of five, most children have developed the neural capacity to distinguish between reality and fantasy, so six-year olds continue to believe primarily because their parents encourage them to. However, in one intriguing study of children's beliefs concerning the Easter bunny, the researchers found that in families where the parents discouraged such beliefs, 47 percent of the children continued to believe that the Easter bunny was real. In families that encouraged such beliefs, 23 percent of the children chose to disbelieve.[45]

Thus, two factors are at work: parental and societal influence, and the biological capacity to intuit the laws that govern physical objects in the world. In addition, children also have an astonishing ability to discern when their parents are not telling the truth; this is another reason why older children reject the idea that Santa is real. If the parents don't really believe, this is communicated to the child through subtle facial clues.

> I stopped believing in Santa Claus when I was six. Mother took me to see him in a department store and he asked for my autograph.
>
> —Shirley Temple Black

As a child grows older, more discrepancies appear, especially when he or she encounters the infamous department store Santa. For very young children, this stranger—dressed up in an unusual red suit—is far scarier than the Santa in storybooks. The amygdala goes on the alert, and the fight-or-flight response kicks in, as many a Santa Claus's shin has painfully discovered. Other clues alert the child that something is amiss. When my daughter encountered her first Santa in a mall, she immediately exclaimed, "He's got a false beard!" And any child, using the abstractive functions of the brain, can tell the difference between a plastic playhouse and Santa's workshop at the North Pole. Toss in a few fake reindeer and a few bored actors in little green suits, and no one over the age of two will believe in the reality of the scene.

When the child encounters multiple Santas in other department stores (these Santas violate quantitative and causal rules), the belief is further dispelled. But even though the illusion is shattered, the overall concept (which makes use of integrative holistic functions) can still be fun, and so many children will hang onto their fantasies for a few more years.

A belief in God, however, is a different matter. First, the majority of parents who introduce their children to spiritual beliefs usually believe deeply themselves. Furthermore, most parents spend a great deal of time discussing such issues with their children; and even parents who don't believe in God tend to encourage their children to think openly about such concepts.[46] An even more intriguing finding is that children between the ages of seven and nine, who were raised by secular parents, preferred a creationist model concerning the universe over an evolutionary one. Not until they reached adolescence did they embrace the evolutionary beliefs of their parents.[47] This adds credence to the hypothesis that the human brain is inclined to accept the reality of spiritual beliefs, separate from the influence of others. In a related study, researchers found evidence that different types of hemispheric interactions in the brain might bias an individual toward embracing or rejecting creationist and evolutionary perspectives.[48]

God and the Power of the Unseen

I sometimes wonder if the belief in Santa Claus would stay alive longer in children if the jolly old man were never seen in the flesh. Invisibility and ambiguity, coupled with an acceptance of reality from the community at large, seem to be essential elements for maintaining such beliefs. From time immemorial, sea monsters were believed to be real, in part because ships disappeared in unexplainable ways, and because no one had actually seen a monster. As long as you don't encounter visual evidence to the contrary, superstitious beliefs can continue without interruption until a satisfying alternative explanation is embraced. In the past, magical thinking was considered a respectable philosophy, since it provided adequate explanations for mysterious events. Early scientists, like Isaac Newton and the majority of the Royal Academy, vigilantly maintained their interest in magic and alchemy,[49] topics that were supported—or at least tolerated—by the church at that time because they appeared to capture some truth.

This notion of the unseen is an essential component in the vast majority of religious traditions. When you can grasp something—like God—with your everyday senses, you cannot dissect it in any ordinary way. Thus you cannot quantify God or reduce God into separate parts. Nor can God, at least according to the ancient Hebrew texts, be named. But to refrain from naming God interferes with the abstractive function that is designed to label every object in the world. If you can't perceive something, then how can it exist? A young child might say, "It just does," and a priest might say, "You need faith."

But the brain will still strive to solve such paradoxical thoughts by creating rational explanations of how such realms can exist. The Vatican library, for example, contains thousands of treatises that have attempted to categorize, label, and reduce the mysteries of the spiritual realm to fit our everyday understanding of the world. We introduce dualistic and oppositional notions such as heaven and hell, or good and evil, and we attempt to impose causal reasoning on how they will affect our lives. For example, some people come to believe that if you pray with sincerity, God will listen and respond to your concern. But if God works in mysterious ways, as most religious texts imply, how can we ever know? And so we search for miracles, or for other evidence or proof, because it is far more difficult, at least as far as the normal functioning of the brain is concerned, to simply have faith in the unknown.

As long as God remains a mysterious concept, seekers will be drawn to what they do not understand, questioning and imagining what the reality or truth might be. In this sense, religious believers struggle with God in much the same way that physicists struggle with quantum mechanics, or a teenager struggles with love. Unfortunately, every attempt to grasp the totality of life from a cognitive perspective is bound to fall short, because the brain is limited in how it perceives the world. We may try to fill in the gaps with intricate beliefs, but ambiguities and uncertainties remain, and these are the very things that pull us deeper into our biological quest for knowledge, meaning, and truth. Around and around we go, for wherever there is a mystery, our brains are destined to explore.

Part II

Childhood Development and Morality

Chapter 5

Parents, Peas, and "Putty Tats": The Development of Childhood Beliefs

SOME OF MY DEEPEST BELIEFS ABOUT LIFE CAME FROM A conversation I once had with my mother concerning peas. As I recall, I was about four years old, sitting at the dining room table, and like many kids, I loathed the flavor and texture of peas. I knew that it was important to eat my vegetables—my parents were relentless on this topic—but for the life of me, I did not believe those lifeless pellets could possibly be good to eat. My mother had tried countless methods to get me to finish my peas, but my stubbornness usually won out. As any parent knows, a four-year old can tenaciously cling to any preferential belief, particularly regarding peas.

One day she tried a different approach: guilt. She told me that the peas on the plate would be lonely if I left them there because they wouldn't be with their friends. Their friends, of course, were the peas I had already reluctantly eaten. Suddenly, I saw my plate in an entirely different light. Peas, I realized, had feelings—and friends!

"How would you feel," my mother insisted, "if you lost some of your friends?" Without hesitating, I quickly swallowed the rest of my peas.

Although I didn't realize it at the time, I had been primed for an entirely new understanding of the world. From that moment on, I tended to view virtually all objects as animate, with feelings and thoughts. I also began to believe that everything was somehow fundamentally connected. Whether it was the food I ate, or my family and friends, I felt that we all were bound to each other by some unseen mechanism or force.

Even my toys took on a semblance of life. I wouldn't leave my baseball glove on the porch at night, because I believed it would be frightened if it was left out in the dark, alone. And when I played with my building blocks, I had to use every single piece for fear of hurting the feelings of the unused ones. After all, no one likes to be excluded from a game.

These childhood beliefs would soon influence my social behavior at school, for I couldn't stand it when one of the other kids would be excluded from our games. I would insist that he or she be included, and if the team refused, I'd quit. Even in my high school debates, I often had trouble discarding the opponent's view. You could say that my experience with peas made it easier for me to swallow—or at least sample—other people's opinions and beliefs.

Today, I continue to treat the world as an interconnected whole, where everyone and everything has value and a place and is to be treated with equal kindness and respect. And it all began with a series of childhood beliefs: the belief that you have to listen to your parents (at least some of the time); the belief that vegetables are good for you (at least some of the time); and the belief—which was triggered by an overwhelming feeling of guilt—that peas had feelings and friends.

Powerful emotions create strong memories; and memories, when coupled with language, are the basis for forming conscious beliefs. This level of belief is what we often call "knowledge,"[1] but if it doesn't have an emotional appeal, the belief will not register deeply in a person's mind.

The emotional experience associated with eating peas induced one of the earliest moral beliefs of my childhood: peas are fundamentally evil. The fact that somebody else tells you they are good for you is not enough to override the emotional desire to throw them across the room, especially since a child does not have the cognitive capacity to appreciate the concept of nutritional health. But the concepts of friends and loneliness meant a great deal to me, and it encouraged me to suspend my gut reaction to peas.

Language, Gorillas, and Beliefs

Among preadolescents, many beliefs appear to be based primarily on the emotional impact of specific experiences. This is true for other primates as well as for humans. Gorillas and chimpanzees use various gestures and facial expressions to communicate with each other in the wild. Koko, the world's most famous "talking" gorilla, has been taught to use sign language to communicate. When Koko lost her pet kitten in a car accident, she became distraught. Later, when she was shown a picture of a similar-looking cat, Koko pointed to the picture and signed, "Cry, sad, frown." Then, when she was asked what she thought had happened to the kitten, Koko signed, "Sleep cat," expressing the same type of belief that young

"I'm a Good Gorilla"

To express this belief about herself, Koko uses American Sign Language. In a video clip, which you can watch at http://www.koko.org/, she first pats her left shoulder with her right hand, a gesture that means "Koko." Then she points to her chest. Next, she gestures with her left hand in front of her mouth, signing the word "good." Koko also refers to herself as a "fine animal person gorilla."

children have about death.[2] Koko understands 2,000 words in English and has learned how to use 1,000 sign-language gestures to communicate.

Koko, by the way, believes that she is a good gorilla. In humans, this moral principle begins to develop around the age of eight. Koko likes romantic movies and video dating, and she wants a baby of her own. According to Francine Patterson of Santa Clara University, who "adopted" Koko in 1972, gorillas have the capacity to exhibit nearly every aspect of human language, communication, and consciousness—in other words, all the criteria essential for the formulation of conscious beliefs:

> In addition to intensive studies of vocabulary acquisition, the project has investigated spontaneous gorilla language use [involving] the study of innovative linguistic strategies, invention of new signs and compound words, simultaneous signing, self-directed signing, displacement, prevarication, reference to time and emotional states, gestural modulation, metaphorical word use, humor, definition, argument, insult, threat, fantasy play, storytelling and moral judgment.

Koko and her gorilla partner, Michael, also like to paint abstract pictures of birds, people, and emotions such as hate and love. "It is part of ape nature to paint," says the primatologist Roger Fouts of Central Washington University. "Apes like to use crayons, pencils, and finger paints. Of course, they also like to eat them."

We humans like to believe that we are smarter than other living creatures, and we often cite our language skills as proof. But before you jump to any conclusions, consider this: a gorilla can learn how to communicate to people using rules that govern human speech, but we have yet to learn how to have a conversation with gorillas using their forms of communication.

Childhood Memories of Events That Never Occurred

Because many neuronal connections are still unformed, a child's perceptual and cognitive evaluations of people's moods and feelings are often different from those of an adult. Also, childhood memories and beliefs turn out to be particularly inaccurate and can be easily influenced—even falsified—by other people. But because they've been repeated and reinforced over many years, those memories are often the least likely to be modified or rejected as a result of later experiences and beliefs. Jean Piaget (1896–1980), whose models of childhood development continue to shape contemporary education, tells a story that exemplifies how easy it is to construct false memories and beliefs in childhood:

> I can still see, most clearly, the following scene, in which I believed until I was about fifteen. I was sitting in my [baby carriage], which my nurse was pushing in the Champs-Elysées, when a man tried to kidnap me. I was held in by the strap fastened around me while my nurse bravely tried to stand between me and the thief. She received various scratches, and I can still see vaguely those on her face. Then a crowd gathered, a policeman with a short cloak and a white baton came up, and the man took to his heels. I can still see the whole scene, and can even place it near the tube station. When I was about fifteen, my parents received a letter from my former nurse saying that . . . she had made up the whole story.[3]

Piaget realized that he must have heard the tale from one of his relatives "and projected it into the past in the form of a visual memory." I, too, distorted the memory of my mother and the peas, for I used to believe that I was three years old when the event took place. My mother doesn't recall it at all. There is even a possibility that I made up some parts of it. Nevertheless, it remains one of the most potent memories of my childhood.

Current research suggests that many of our memories about ourselves—particularly those about our early years—are partly a form of wishful thinking, an internal cognitive process that attempts to restructure our autobiography in a positive light.[4] In other words, each time we recall an old memory, we tend to deemphasize its negative aspects while highlighting, and often embellishing, the positive aspects. According to the distinguished professors Reid Hastie and Robyn Dawes:

> We quite literally make up stories about our lives, the world, and reality in general. The fit between our memories and the stories enhances our belief in them. Often, however, it is the story that creates the memory, rather than vice versa.[5]

I Tawt I Taw a Putty Tat—at Disneyland!

The brain's cognitive and emotional processes enable us to form memories that establish the beliefs we will carry throughout our lives. Memories are essential for understanding the world because we must constantly compare current perceptions with previous perceptions. Yet many of those memories have no clear basis in reality.

False memories, and the beliefs they generate, are common, and memories can be fabricated in controlled laboratory situations. One eminent psychologist, Elizabeth Loftus, has conducted with the aid of her students more than 200 experiments involving more than 20,000 individuals that document how easy it is to create false memories by feeding misinformation to the subjects. "Give us a dozen healthy memories, well-formed, and our own specified world to handle them in," writes Dr. Loftus, "and we'll guarantee to take any one at random and train it to become any type of memory that we might select . . . regardless of its origin or the brain that holds it."[6]

Dr. Loftus's early studies focused on the inaccuracy of eyewitness reports of accidents and crimes. She demonstrated that a person's belief in what was seen could be changed by merely shifting the emotional content of a question. When adults were asked, "How fast were the cars going when they *smashed* into each other?" they estimated higher speeds than when the word "hit" was substituted for "smashed."[7] The memory of a witness, then, can be altered, depending on how a question is asked.

A large body of research now supports Loftus's original work—which showed how emotions, ambiguous situations, and misinformation obtained through gossip, rumor, and hearsay could generate a variety of false beliefs, especially beliefs concerning childhood experiences. Researchers have implanted false memories about being lost for long periods of time, of being in life-threatening situations, of molestation and rape, and of ritual abuse and demonic possession.

People have even been led to believe that they were born left-handed.[8]

In one of Loftus's studies, more than one-third of the adult subjects were tricked into believing that they had interacted with Bugs Bunny during a childhood visit to Disneyland. Half of those believed that they had shaken hands with him or hugged him, and more than one-fourth "remembered" touching his ear or his tail, or hearing him say, "What's up, doc?" These beliefs were utterly false: Bugs Bunny is a Warner Brothers' creation and does not appear at Disneyland. As one staff writer for the *Los Angeles Times* put it, if Bugs were to show up at Disneyland, "the wascally . . . wabbit would be awwested on site."[9]

Loftus's work has also helped to improve our criminal justice system, and for this reason she was named in the *Review of General Psychology* as one of the most important psychologists of the twentieth century. Judges, lawyers, and juries have become aware of the inaccuracies inherent in memory, and as a result changes have been made in how police interrogations are carried out and how evidence can be collected and used. Loftus's work has also helped others to expose and overturn many wrongful convictions.

The Problem with "Recovered" Memories

In the game "Twister," the players compete to place their hands and feet on a large sheet covered with circles of different colors. In a matter of minutes, the players are usually awkwardly entwined and falling over each other in laughter and embarrassment. Researchers at State University of New York recently used this game to explore how children ages four to seven would react when they were asked some sexually provocative "leading" questions a week later.[10] The researchers got some shocking responses, for—depending on how they put the questions—the answers began to sound like sexual abuse. If they simply asked, "Did Amy touch your bottom or kiss you?" the children would usually answer no. But when the question was asked in a different way, more children said yes. Can you guess how the question was rephrased? This is what provoked a higher affirmative response: "Amy kissed you, didn't she?" "She touched your bottom, didn't she?" All it took to "recover" a false memory from a child was the indirect assumption that a particular event actually occurred.

The implications of this study, and others like it, are clear: if it is so easy to manipulate and alter the beliefs of children concerning acts of sexuality, how do we discern whether adult memories of childhood sexual abuse are accurate or not? This issue made national headlines in the 1980s, when hundreds of cases were filed by purported victims who believed that they had been abused in childhood. The plaintiffs claimed that the memories had been repressed for years or decades because they were so painful, but that with the help of a therapist, they had been able to "recover" these lost memories. The memories were elicited using techniques like hypnosis and psychodrama, and the therapists were convinced that the recovered memories were real. Many talk shows on television featured the presumed victims, whose stories encouraged more people to "recover" their own memories and file more cases in court. Across the country, hundreds of self-help support groups were formed to help victims "realize" that their fears and fantasies were true.

As these cases came to trial, however, evidence rapidly accumulated showing that many of the recalled memories were false, or at best unsubstantiated. Then, a number of the accusers recanted their stories, claiming they had been coerced by their therapists into believing that the recovered memories were real. Then came a flood of lawsuits, this time against the therapists, some of whom went to jail. In 1995, in an article entitled "Remembering Dangerously," Elizabeth Loftus described a series of cases in which undercover investigators went to licensed therapists' offices pretending to be patients[11]:

> In one case, the pseudopatient visited the therapist complaining about nightmares and trouble sleeping. On the third visit to the therapist, the investigator was told that she was an incest survivor.[12] In another case, Cable News Network[13] sent an employee undercover to the offices of an Ohio psychotherapist (who was supervised by a psychologist) wired with a hidden video camera. The pseudopatient complained of feeling depressed and having recent relationship problems with her hus-

band. In the first session, the therapist diagnosed "incest survivor," telling the pseudopatient she was a "classic case." When the pseudopatient returned for her second session, puzzled about her lack of memory, the therapist told her that her reaction was typical and that she had repressed the memory because the trauma was so awful.

Recovered-memory therapy has been shown to be unverifiable, and memories recovered in this way are no longer admissible in court.[14] In a recent study, victims claiming recovery of repressed memories were found to be more prone to fantasies and false recollections of other remembered events.[15] Subsequent research has even found that interpretation of dreams can be used to alter a person's memories and create false beliefs.[16] It has also been repeatedly proved that coercion—by a therapist; an interrogator; or, for that matter, any authority figure—will strengthen a person's belief in the validity of any recalled memory or event.[17]

In early 2002, the controversy about recovered memories surfaced again when numerous children and adults claimed to have been molested by priests. Once again, it is difficult to determine how many cases are real, and how many have been unconsciously fabricated owing to authoritarian pressure and the ensuing media frenzy. Sadly, when not enough evidence is found, some perpetrators will probably go free; on the other hand, thanks to psychological research, most innocent defendants will not end up in jail.

Building Beliefs on a Wing and a Prayer

We are all prone to false memories and beliefs. A United States senator once described a political speech that had deeply moved him as a child; unfortunately, a member of the press pointed out that the senator had not been born at the time the speech was given.[18] Even presidents are vulnerable to confabulation. During the campaign of 1980, Ronald Reagan would often cite a story about World War II in which a pilot whose plane has been hit by enemy fire is going to jump out and parachute to safety. One of his crew, however, is too badly injured to jump. "Never mind," the pilot tells him, "we'll ride it down together." Reagan believed that this was a true story, and he would have tears in his eyes as he recited it, but it turned out to be a scene from the movie *A Wing and a Prayer* (1944). "Reagan had apparently retained the facts but forgotten their source," says Daniel Schacter of Harvard. This is the kind of mistake that nearly everyone makes when recalling events from the past.[19] Again, we see how the power of emotion can turn fantasy into a supposed fact. Schacter also emphasizes that memory itself is a specific kind of belief about the past, one that is particularly difficult to substantiate or prove.[20]

Memories can also be scrambled and confused, and they often include mistaken details. These erroneous bits of misinformation are themselves encoded into new memories, which are later recalled and altered again, replacing the older, more accurate perceptions. It even turns out that false memories are more difficult to dismiss,[21] perhaps because the dissonance between fact and fiction causes a stronger emotional reaction within the limbic areas, which in turn interferes with our ability to use logic and reason in evaluating our beliefs about the world.[22] Additionally, the more traumatic an event, the more likely the victim is to construct beliefs that border on the bizarre.[23]

The addition of visual suggestions and cues can also make it more difficult to distinguish between true and false beliefs. This has serious implications for criminal law, because a victim who is shown photographs of possible perpetrators is more likely to believe that one of them was the instigator of the crime.[24] Furthermore, the more visual the imagery, the more neurally inclined we are to believe that a particular memory is true.[25] In one experiment, a fake image was constructed by superimposing an actual photograph of a child and father onto a picture of passengers in a hot-air balloon. When the altered composite photograph was shown to test subjects, half of them believed that they had actually flown in the balloon.[26]

The university researchers called their paper "A Picture Is Worth a Thousand Lies"; but in a later study they discovered that subjects were more inclined to believe a fabricated narrative account about being in the balloon.[27] They titled the later paper, "Actually, a Picture Is Worth Less than 45 Words." In either case, the lesson is obvious: when we read, see, or hear autobiographical accounts—whether they are about presidents, CEOs, or our own parents—it is wise to be skeptical regarding their accuracy.*

The Neurobiology of Memories and Beliefs

Neurological disorders can also interfere with our ability to distinguish between accurate and inaccurate memories. Brain imaging studies suggest that the right prefrontal cortex plays a crucial role in integrating current perceptions, ideas, and memories, and that abnormalities within this integrative process can cause strange and unusual beliefs.[28] When the limbic system is damaged, a patient can lose the ability to suppress fantasies that do not pertain to ongoing reality.[29]

Imaginary memories and reality-based memories are stored in different parts of the brain,[30] and if the neural pathways between these areas are interfered with, a person may lose the ability to distinguish between fantasies and facts. For instance, common antianxiety drugs such as Ativan, Valium, and Xanax can impair the conscious recollections of true memories but not false memories.[31] The reason for this, presumably, is that accurate memories require a high degree of neural organization, which can be disrupted by drugs. Other studies have found that antianxiety drugs disrupt both true and false memories by causing us to exaggerate "the personal significance and emotional intensity of past events."[32]

Memories are also affected by stress. Scientists at Yale concluded that the neuropeptides and neurotransmitters released during stress can alter the functioning of areas of the brain directly involved with memory formation and recall.[33] "Such release," the researchers wrote, "may interfere with the laying down of memory traces for incidents of childhood abuse," and possibly lead to long-term distortions of the facts, or even amnesia. To complicate matters, highly emotional and traumatic experiences seem to enhance memory storage, but in a fragmented way, subject to temporal and spatial distortions.[34] To compensate for the inconsistencies inherent in memory recall, the brain constructs alternative scenarios as it rebuilds a coherent worldview.

To date, the accumulated research pertaining to the accuracy of our memories and beliefs can be summarized as follows:

- All memories and beliefs are subject to change and distortion over time.
- Conscious beliefs and memory recall are highly dependent on language, emotion, and social interaction; as these variables change, so do our memories and beliefs.
- Children's memories and beliefs distinguish poorly between fantasies and facts.
- The older a memory, the more difficult it is to ascertain accuracy.
- Autobiographical memories are particularly prone to inaccuracy.
- Traumatic events embed memories in a powerful but somewhat fragmentary way.
- Neurological disorders and drugs can disrupt the brain's ability to distinguish between true and false memories and beliefs.

There is one more essential component in the formation of beliefs: belief systems tend to develop in ways that parallel the brain's own biological development and the acquisition of knowledge and social skills. So, since the information we acquire and the ways our brains interpret and integrate it neurologically are unique for each person, the belief systems we end up with can be highly individualistic.

Morality and Childhood Development

So far, our discussion of beliefs has focused primarily on the perceptions, feelings, and thoughts that relate to objects and events outside of ourselves. Another level of belief formation involves the way we relate to people and the values we place on these interactions. As children learn which behaviors work best in different social situations, they begin to learn the concepts of good and bad, right and wrong, and fairness and unfairness, and assemble them into a rudimentary

Piaget's Stages of Cognitive Development		
Birth to age 2	Sensorimotor	Infants construct their sense of reality by coordinating perceptions and feelings with body movements. This allows them to believe that objects exist permanently in the world.
Ages 2 to 7	Preoperational	Children begin to represent the world symbolically, with words, images, and drawings. They are prone to fantasies about reality and are primarily self-centered.
Ages 7 to 11	Concrete operations	Concrete ideas about reality (size, shape, etc.) replace fantasies and intuitive thoughts. Children can also begin to test simple hypotheses in systematic ways.
Starting at age 11 or 12	Formal operations	Adolescents think more logically, abstractly, and idealistically. They are preoccupied with relationship dynamics, but remain largely self-centered.

system of moral beliefs. They slowly begin to distinguish between actions that foster a sense of well-being in others and actions that cause harm. This distinction is essential if they are to develop so that they can cooperate with other individuals and groups in society.[35] This moral development is actually embedded in our genes.[36]

Jean Piaget's childhood stages of cognitive development are also useful in demonstrating how morality develops. Piaget found that a child builds a progression of logical beliefs about the world that correspond with the stages of maturation of the human brain.[37] The table above summarizes Piaget's four-stage model.

In the 1960s, David Elkind, through his studies of children from different faith traditions,[38] demonstrated that religious thinking goes through a similar development in stages. And in the 1970s, Lawrence Kohlberg expanded Piaget's model to include six stages that cover the emergence of beliefs about self-worth, sharing, loyalty, justice, punishment, fairness, and morality. Over the past thirty years, Kohlberg's model has been supported by numerous cross-cultural studies.[39] Nonetheless, these theories about how the moral and spiritual beliefs of adolescents and adults are shaped are still controversial. Yet I think that a child's belief systems—including religious and spiritual beliefs—clearly coincide with the neural development of the brain. The next sections of this chapter will trace this evolution.[40]

Stage I: Infancy and the Absence of Belief

As far as we can tell, newborns have no discernible beliefs, because the newborn brain is at a primitive stage of development, barely able to integrate sensory information.[41] Basically, then, infants are amoral because their brain cannot differentiate between right and wrong, concepts that depend on the functioning of a more mature brain. In fact, a neonate's brain metabolism is 30 percent lower than that of an adult.[42]

At this early stage, brain development is highly dependent on the quality of care the infant receives. When there is inconsistent care or neglect, the infant's brain functioning may be impaired. For example, animal research has found that if the mother-infant relationship is disrupted, long-term neural changes occur that can lead to "increased vulnerability to aging and to psychopathology"[43]—in other words, to mental and physical illnesses. And if the mother is separated from an infant for too long, the baby's brain function and development are disturbed.[44]

Intense insecurity in human infants can later inhibit the child's ability to cope with stress.[45] A lack of connection with caregivers leads to a lack of interconnections between brain cells—a lack of organization within the brain. Studies demonstrate that a lack of proper care will hinder a child's higher cognitive functioning, preventing the development of adequate social skills.[46] However, it appears that later intervention (compassionate nurturing, environmental enrichment, etc.) can reverse the damage caused by the earlier stress.[47]

By contrast, the brain of an infant who is raised in a healthy, nurturing, and stimulating environment will show increased neural activity and maturity by the end of the third month.[48] Theoretically, such a child will feel less anxiety and depression in later life, and exhibit greater social skills with others.

The 10 Percent Myth

The belief that we only use 10 percent of the brain is a myth, or what I like to call a "friend of a friend" belief. Dale Carnegie mistakenly credited it to William James. Others have wrongly attributed it to Einstein. Some have tried to substantiate this popular misconception by citing excess or silent neurons, or the fact that 90 percent of the brain is composed of glial cells. But glial cells are also essential for neural activity. The fact of the matter is this: every cell in the brain is alive and is actively stimulating other parts at various times throughout the day. We may lose half of the brain cells by the time we reach age thirty, but not because they are unused. They simply are not needed. In truth, too many neurons hinder the functioning of the brain.

Stage 2: Learning How to Play by the "Rules"

The next stage of belief development occurs roughly between the ages of one year and six years. Children begin to use symbols, language, imagination, and speech to organize their sensory experiences into a meaningful, though primitive, system of beliefs. They can understand the concept of cause and effect, but are generally loose about applying it in a logical way. One can see this easily in the game-playing strategies of a four-year-old. When my daughter plays checkers with her friends, they ignore nearly all the rules I try to impose. I'll tell them how the pieces are supposed to be moved, but they will place pieces anywhere they want, and they take great pleasure in removing pieces whenever they feel inclined. Nor do they care about winning.

Developmentally, young children don't have much of a clue about how anything works—games, friendships, siblings, or life in general. They're just learning how to get along with others, and as they master the rules of simple games, they begin to establish basic cooperative beliefs. When my daughter turned five, she decided to follow the rules. Now, when we open up the box of checkers, she'll proudly proclaim, "It's more fun when everyone agrees to take turns."

At the California Institute of Technology in Pasadena, California, John Allman and his colleagues identified a unique structure in the brain called the spindle cell, which appears around the age of four months and gradually increases in quantity and size in the first three years of life.[49] Spindle cells have been found only in primates and humans, and appear to be linked to our ability to develop a moral sense. According to Dr. Allman, they exist in a mysterious part of the brain (the frontoinsular cortex) and are activated when a person perceives unfairness or deception.[50] The fact that it takes many years for these unique cells to develop helps to explain why children younger than four are unable to play games fairly or follow the rules.

For children, concepts involving right and wrong are more difficult to grasp than those involving good and bad. A three-year-old might say, "It's bad to steal," but when asked why, she'll simply reply, "Because you'll get punished." The child knows how bad it feels to be reprimanded but doesn't understand the effects that stealing has on others. One time, when some friends were visiting my family, their little girl spied a large bowl of sweets that I put out on the table for such occasions. Timidly, she asked her mom if she could have some candy, and got permission. Later, when we came back into the kitchen, the bowl was empty, but the little girl's pockets were bulging at the seams. She looked very happy, and did not attempt to put back the candy until her mother told her to do so. The child knew that it was "right" to ask permission, but she didn't know how much it would be fair to take. Her logic was weak; her moral sense was only beginning.

At this stage of development, children do start to make moral appraisals, but the circuits that govern moral awareness take many years to mature. If the brain is injured during this time, the child may lose the ability to respond effectively to emotional and social cues.[51]

Young children's brains have an overproduction of neurons throughout, especially in the frontal lobes[52]—the area responsible for logic, reason, and conscious control. This overconnectedness has been associated with an increase in fantasy that gives rise to belief in monsters, fairies, and a host of other imaginative creatures. Children have little ability to discriminate between reality, fantasies, and dreams. In one study, researchers discovered that four-year-olds who were shown pictures often made up additional information when asked about what they saw, but significantly fewer errors of confabulation occurred in older children. The researchers surmised that this occurred because there was greater separation between the two hemispheres of the older children's brains.[53]

In other words, younger children have too many connections, which generate false information about what they see. In addition, emotional issues further complicate and distort a child's view of reality. My daughter once complained that one of her classmates, Sandra, was hitting her. We talked about how she might handle the situation in order to avoid getting hit again, but it didn't work; my daughter said that Sandra kept hitting her. So I went to the school to talk the situation over with the teacher. It turned out that Sandra had moved away two months earlier. When I pointed this out to my daughter, she burst into tears, confessing that she was mad that her friend had moved away. In a young child's mind, being hit and being hurt feel the same emotionally, and for my daughter, this was enough to alter her realistic sense of the world.

Hemispheric connections do not become fully mature until a child is about age ten.[54] Over time, these excess neural connections will slowly be pared away as the brain decides which neural circuits are the most useful for survival. During the pruning process, most children will develop more realistic views and give up many, though not all, of their magical beliefs. Until then, they will make use of a very important tool to help them organize their feelings and thoughts about the world: storytelling. The most important stories are those that incorporate cultural and religious myths. By identifying with the characters in these stories, young children vicariously experience moral conflicts and solutions that will have great relevance later in life. And, as dozens of studies have shown, adult's belief systems—especially those concerning religion and spirituality—will contain significant remnants of the stories these adults heard and read while growing up.[55]

Along with a conscious recognition of selfhood, young children also come to realize that they can die. This realization can be very frightening, and a belief in an afterlife can be comforting. Belief in an afterlife is often introduced to a child through religious stories and mythical tales that may describe wondrous worlds existing beyond the boundaries of life. In Asian mythology, there are heavens and hells and ghosts of every imaginable kind, and variations of reincarnation. In Australian Aboriginal myths, people transform themselves into mountains, rivers, and trees. As the thirteenth-century poet Rumi wrote:

As I died from a mineral, I became a plant,
And from the plant I rose to animal.
I died from an animal and became a man.
Why, then, should I fear disappearance through death?[56]

Young children are deeply affected by such stories and poems, which often form the foundations of their religious and spiritual beliefs. But since their brains have not yet formed consistent associations between causes and effects, some amusing conceptualizations can occur. For instance, after my neighbor's father died, his daughter asked where Grandpa was. She had recently had a difficult time with the death of her pet goldfish, so in order to avoid any further upset, her father informed her that Grandpa had been taken to the cemetery and planted in the ground. This turned out to be an unfortunate choice of words because one day, while my neighbor was out tending his roses, his daughter ran up to him, extremely worried, and blurted out, "We've got to go to the cemetery now, and water Grandpa before he dies!"

The spiritual realm fascinates people of all ages; but children between the ages of three and eight try to envision it in concrete terms, and they usually turn spiritual deities into human forms with superhuman powers.[57] Interestingly, Jewish children represent God more symbolically and abstractly than Christian children, who usually give God a face.[58] In children's conversations, God also tends to take on a strong authoritarian personality—sometimes kind, sometimes punitive. In this manner, psychologists argue, God becomes an internalized figure of parental authority and the basis of one's conscience,[59] and our first images of God are constructed through our parents.[60]

As children start to conceive of God, they are also struggling to develop a working concept of right and wrong, good and bad, an essential step toward developing socially accepted behavior. In this sense, religious teachings help children to establish their fundamental moral beliefs.

But because their cognitive functions are not fully operational, young children mix logic with fantasies and fears: "If God exists, where does he live? Is he watching me now? Does he know I hit my sister when Mom wasn't looking?" One child, who took his parents literally when they told him that God can see everyone, everywhere, and at any time, refused for months to bathe without a swimsuit. This exemplifies how feelings of guilt and shame, which are common at this stage of development, become infused with religious concepts and beliefs.

Many young children cannot tell if they will be punished or praised for their behavior by an unseen, omnipotent God. Some parents intuitively recognize this power and thus may be inclined to invoke a fear of God's wrath as they try to encourage their children to behave in specific ways. However, there may be a price to pay for taking this approach. According to extensive research carried out by two psychology professors—Bob Altemeyer and Bruce Hunsberger—children who grow up in fundamentalist families do tend to obey the authorities and follow the rules, but they also tend to be self-righteous, prejudicial, and condemnatory toward people outside their groups.[61] They have an "us versus them" mentality that many will carry throughout their lives. On the other hand, fundamentalist congregations experience a 50 percent dropout rate among their members.

Imagine being a child in 1741 in church on the day Jonathan Edwards, the American Puritan preacher, delivered one of the most famous sermons in American history. Edwards shouted from his pulpit, for nearly an hour, that everyone in the room was a sinner, teetering on the edge of a fiery pit: "And you, children, who are unconverted, do not you know that you are going down to hell, to bear the dreadful wrath of that God, who is now angry with you every day and every night?" You might as well read Stephen King to your four-year-old at bedtime.

A punitive god, like a punitive parent, encourages children to internalize anxious and potentially destructive concepts. As negative beliefs develop, they too become embedded in the neuronal connections being formed in the brain, and this makes them difficult to relinquish later on in life. Fortunately, the majority of religious groups successfully instill a disposition to forgive others.[62] Belief in a compassionate and forgiving God can give a child a sense of optimism and safety.

Stage 3: Learning How to Play Fair

Between the ages of six and ten, children become more and more concerned about separating fantasy from fact. They also pay closer attention to personal exchanges and to stories that relate to membership in a community or group. This is an essential step toward developing personal beliefs within a social context.

At this age, children begin to realize that different authorities have different beliefs and rules. "Who, then, should I believe?" the child wonders. "My mother, the teacher, or the priest?" Instead of just deferring to the adult with the most power, as a younger child would, older children will learn how to form alliances with those people who they think are most likely to satisfy their needs. In this manner, children learn to share their toys, not necessarily because it is the "right thing to do," but because they have figured out that if they give a toy to a friend, they can take one in exchange, without the friend's getting mad. The lesson here in the benefits of cooperation is obvious: "You scratch my back and I'll scratch yours."

Relationship exchanges are also based on bonds of loyalty and gratitude, and when somebody does a child a big favor, the child is more inclined to reciprocate. Since children at this stage are primarily driven by narcissistic impulses and desires, they still have difficulty embracing fully notions such as equality, fairness, and justice. But you can see the beginnings of such beliefs creeping into their consciousness. A six-year-old, when asked to put away the dishes, might complain bitterly, "No, Mom, my favorite TV program is about to begin." An eight-year-old, however, will often introduce a fuzzy notion of fairness: "Why do I have to put away the dishes every night, while my sister gets to play?" You'll reply, "She's not playing, she's doing her homework," but it will take less than a second before he raises another objection: "Come on, Mom, you're not being fair. All she has to do is take out the garbage, and she doesn't even do that every day."

Although a sense of fairness is emerging, children at this stage are still governed by selfish impulses. They may behave morally for fear of punishment, but if the opportunity arises, they will relegate morality to the backseat and let self-interest drive them. Here the logic is simple: don't get caught.

The older kids get, the more logical they become, and the more they believe in the validity of their thoughts. They express a belief in absolute terms, but they do not have much ability to step back and question its accuracy. They can also become trapped in all-or-none thinking, for instance, and come to believe that they, or others, are irredeemably bad or good. This type of moral absolutism can last for years, but it often goes unnoticed, since children do not have strong communication skills concerning feelings and moods.

In a conversation with a group of nine-year-olds, Robert Coles, a professor at Harvard, discovered that fights can break out as children grapple with their conceptions of God. One boy compared God to a doctor, "sitting and trying to be a friend, and maybe praying." This angered one of the girls, who retorted that God doesn't pray: "God is God, so why should he pray to himself?" Another boy fired back, "How do you know what he does? Did you talk to him? Why couldn't—why wouldn't—he want to pray?" The boy then cautioned the others not to believe what leaders say about God: "Anyway, my dad says God isn't owned by ministers and priests and rabbis." Another girl interrupted, "That's your father's idea, but it's not my dad's!"[63] As this brief exchange shows, young children often form emotionally inflexible ideas about God, primarily based on their parents' beliefs. At this age, they do not show much tolerance or compassion for those who hold different beliefs.

During this period, the brain continues to "prune" or cut down on connections between nerve cells,[64] a process that is essential for the development of complex systems of logic. Useless ideas, and the neural circuits that support them, are being destroyed while the neural connections that support important beliefs are strengthened. For example, when a six-year old is learning math, dozens of neural circuits form as the brain tries to figure out what 1 + 1 means. Is it 2, or 3, or 11? The circuits that come up with incorrect answers are pruned or cut away, and the information that remains becomes established as truth: 1 + 1 = 2, and that's it. But more complex beliefs, including social, political, and religious beliefs, require many interconnections, and thus the pruning process has a dramatic effect on a person's entire way of thinking about the world.

Stage 4: Adolescence, Chaos, and Creativity

Adolescence is filled with emotional conflict as children struggle to establish and maintain their own set of beliefs and integrate themselves into society. The World Health Organization defines adolescence as the period of life between ten and twenty years of age. Some researchers, taking into account the various biological, psychological, and sociological changes of adolescence, extend it into the mid-twenties.

Neurophysiologically, adolescence corresponds to a time in which the overall metabolism in the brain begins to decrease. Neural pruning continues (and will continue until metabolic stability is reached at around age thirty). Cognitive processes become stabilized, and the remaining neural connections will govern our thinking and behavior for the rest of our lives. The ability of the brain to create new connections and adapt to new situations decreases notably during this phase.[65] This may seem shocking at first, because it suggests that our ability to grow intellectually becomes significantly hindered as we mature beyond adolescence; but there are ways, which I will describe later, to encourage creativity and openness of thought into adulthood.[66]

Beginning at around age ten, children gradually shift from pleasing themselves toward pleasing others. They seek social approval and want to be part of a group, and so they must learn how to live up to each other's expectations. In this move toward conformity, adolescents' personal beliefs often mirror those of their friends. Language and communication skills also increase, allowing adolescents to discuss their problems with words rather than fists. New ideas can and do emerge, but for the most part, old concepts are simply being adjusted and revised. Still, the young adult can now detect logical inconsistencies in his or her own system of beliefs.

During adolescence, a person's basic beliefs about life, relationships, and spirituality gradually mesh into a coherent worldview, but adolescents continue to struggle between conformity and independence. If they don't conform to the beliefs of their friends, they risk being alienated from the social group. If they do conform, they still must deal with the conflicts that emerge from interactions with members of other groups. In my school, for instance, if you belonged to the math club, you were bound to be ribbed by members of the soccer club. I was spared this "us-versus-them" attitude because I was a member of both clubs. High schools and college fraternities are microcosms of constantly competing groups and subgroups. As conflicts arise, people shift their friendships, seemingly at a whim, but usually to attempt to form more useful social alliances. Failure to resolve these internal and external conflicts can lead to neurological changes that later bring about mood disorders and other psychological ills.[67]

The adolescent characteristically undergoes a highly emotional questioning of personal, social, and religious values. As young adults strive to understand their purpose in life and their place in the universe, they will partially reject parental influences and embrace the beliefs of their peers.[68] But, as you'll recall from earlier chapters, the strength of any belief is a matter of four interacting levels of neural processing: perceptual experiences, cognitive experiences, emotional experiences, and social consensus. In the first year of life, our beliefs are based primarily on our perception of objects in the world and our emotional reaction to them, primarily because these involve the main areas of the brain that are functioning. Brain scans of infants show very little activity in the higher parts of the brain; therefore, there is little cognition and a limited ability to resonate with the behaviors of others. Infants basically believe what they see.

Perception
Emotion
Cognition
Consensus

For an adolescent, social consensus becomes the predominant influence on the process of forming beliefs. And the hormonal effects that are influencing every part of the body combine with feelings to shove aside logic and reason.

Perception
Emotion
Cognition
Consensus

Although this neurobiological formula is simplistic, it gives a sense of why teenage drivers are so dangerous behind the wheel. They're just not paying enough attention to the road. They have a cell phone in one hand and an iPod in the other hand. "Oops," they'll say, "that car came out of nowhere!" A more mature driver emphasizes perception, remaining calm and alert, anticipating (with cognition) potential dangers.

In contrast, a physicist at work engages a very different system of beliefs based primarily on highly abstract cognitive processes. After all, no one can see or touch a quantum property, so the emotional value of the analysis will be limited to the physicist's personal interest and excitement in proving or disproving it. Social consensus is still important, because the physicist's theories will have no meaning if others can't find value in them. But the consensus will be small, since few scientists are investigating the quantum properties of the brain.

Perception
Emotion
Cognition
Consensus

As the adolescent brain completes its neural development, it operates with greater autonomy, which gives rise to a greater array of competing but equally valid beliefs. This may help to explain why religious interests begin to decline at this stage.[69] According to survey data for the years 1976–1996, only 60 percent of American teenagers considered religion an important part of their lives, and only 40 percent reported that they prayed daily and attended weekly religious services.[70] In addition, as Carol Markstrom of West Virginia University points out, only a small minority said that they enjoyed their religious participation.[71] The overall pattern of decline in religiousness during adolescence appears to be reversed among people between the ages of twenty and thirty. During those years there is an increased interest and participation in religion. But thereafter, religiosity tends to decline through the rest of life.[72]

The decline in religiosity during adolescence may not bode well for teenagers' psychological health. Numerous studies have found that religion can help protect an adolescent from depression.[73] In general, religious beliefs seem to offer a variety of benefits, including higher academic performance, higher self-esteem, decreased substance abuse, and greater motivation for volunteerism and civic involvement.[74] Unfortunately, the research literature concerning childhood and adolescent spirituality is sparse, so all these statistics need to be evaluated with care. Furthermore, all the results are related to populations rather than individuals, so any one adolescent may defy these generalities.

Stage 5: Intellectual Maturity and Religious Decline

What happens after adolescence? Piaget's model of cognitive development stops at stage four, but other developmental theorists (for example, Kohlberg, Fowler, Gilligan, Turiel, and Loevinger) have various explanations for the emergence of higher stages of morality, empathy, transcendence, and faith, with beliefs in justice, ethics, and human rights.

Part of the problem with any developmental theory is that once the brain matures, it does not appear to develop further in a linear way. In other words, stages no longer apply. One person goes on to win a Nobel Prize; another takes up permanent residence on the couch, with a handful of chips and a beer. People have choices, and the choices they make cannot be predicted accurately in any statistically meaningful way. To make matters worse, neural development virtually ceases by the time we reach thirty, and it's all downhill from there. The brain's metabolic and neurotransmitter activity begins to decrease,[75] and it continues to decrease throughout the remainder of life (interestingly, the male brain tends to atrophy at a slightly faster rate than the female brain).

Neurologically, enlightenment and peace are unlikely. Even Kohlberg admitted that only a small percentage of adults will reach a moral level at which their lives are governed by higher ethical principles.[76] Nonetheless, this level can be reached by those who choose to work diligently toward the ideals it involves, although this process can take decades of introspection and practice. In later chapters, we'll explore how at any time in life a person—through meditation, prayer, and critical thinking—might be able to transcend the narcissistic confines of adolescence and thereby alter the neural functioning of the brain.

Our brain may lose neural plasticity—the flexibility to change and adapt—to some degree as we age. The decreases in brain metabolism and neurotransmitter activity suggest that as we age, our ability to formulate new ideas and beliefs decreases. A biographical overview of geniuses seems to bear this out. Einstein, for example, published his special theory of relativity when he was twenty-six. Alexander Bell was twenty-eight when he invented his telephone, Edison was thirty when he invented the phonograph, and Francis Crick was thirty-six when he walked into an English pub and announced that he had figured out the structure of DNA. Mozart composed more than 600 works before he died at thirty-five. Bill Gates and Steve Jobs were in their early twenties when they made computer history.

> At eighteen our convictions are hills from which we look; at forty-five they are caves in which we hide.
>
> —F. Scott Fitzgerald

Many geniuses fade with age, but some, like Beethoven, continue to soar. For others, the fruits of their genius will blossom in the later years of life, but their brilliance, if you look carefully, was there from the start. A perfect example is Charles Darwin, who published his great work, *On the Origin of Species,* when he was fifty. However, his insatiable scientific curiosity had filled his life much earlier, between the ages of nine to sixteen. In his autobiography, he writes:

> Looking back as well as I can at my character during my school life . . . I had strong and diversified tastes, much zeal for whatever interested me, and a keen pleasure in understanding any complex subject or thing. I was taught Euclid by a private tutor, and I distinctly remember the intense satisfaction which the clear geometrical proofs gave me. I remember, with equal distinctness, the delight which my uncle gave me by explaining the principle of the vernier of a barometer.[77]

Once the major cognitive functions of the brain become fully operational in mid-adolescence, the stagelike development of childhood is supplanted by multidirectional development, but this is still dependent on the foundation laid by our genes and our childhood experiences. The predispositions are in place so that when we finally find an area of interest, we feel as if it has found us. But we also can choose which areas to develop—our artistic skills, our intellectual skills, our social skills, or our physical prowess. We can head in the direction of business, politics, or the healing arts; but if we want to excel in any specific field, most of us will have to focus our attention on a limited number of goals. We can take up automobile mechanics or brain surgery, but rarely can we be expert at both. A few exceptional people can excel at multiple disciplines, but they will often be deficient in other walks of life.[78] You wouldn't have wanted Einstein to fix your car, or Woody Allen to build an atomic bomb. If Einstein himself had wanted to fix a car, I'm sure he could have learned, but it is doubtful that he would ever have become a great mechanic (likewise, Woody Allen's bombs are limited, thankfully, to the box office).

Nonetheless, as our brain cells decrease, we can continue to build on and strengthen the millions of circuits in the brain. Exercising the brain is like exercising the body: you don't grow new muscles; you just keep the ones you have in better shape. Don't give up trying to excel at whatever interests you most. It keeps the motor running, and it keeps the neurons sharp.

The same holds true for spiritual development: if you want to excel, keep practicing, and you will continue to have strong spiritual beliefs. If you don't keep practicing, you'll probably become more secular. Beginning around the age of thirty (which coincides with the gradual decline in neural efficiency), adults tend to become less religious; and by the time they reach sixty, they pray only half as much and have less certainty (again, a 50 percent decline) that God exists. However, beginning around age fifty, there is an increase in organized religious activities, which provide older people with essential social connections,[79] which in turn, as hundreds of studies have found enhance health and extend life. In addition to social support, higher educa-

> To perceive the world differently, we must be willing to change our belief system, let the past slip away, expand our sense of now, and dissolve the fear in our minds.
>
> —William James

tion, intellectual stimulation, and an optimistic view of one's health forestall the decline of moral reasoning and cognition.[80] And for those who persevere, beliefs may reach a level where moral principles and ideals become the primary force governing personal and social behavior. People like Mother Teresa and Gandhi exemplify this level of moral development, and the way they lived changed entire societies.

Chapter 6

Ordinary Criminals Like You and Me: The Gap between Behavior and Moral Beliefs

IN THE 1950S, CULTURAL ANTHROPOLOGIST COLIN TURNbull went to live with the Pygmies of the Congo, and wrote about life with them in *The Forest People.*[1] He found a gentle people filled with happiness and peace, who treated each other and their environment with a deep respect derived from a strong sense of moral values rarely seen in modern technological societies.

About ten years later, Turnbull wrote another book, *The Mountain People,* which was as disturbing as *The Forest People* was endearing. In *The Mountain People* he described a tribe known as the Ik, who lived in a desolate corner of Uganda where they could barely make ends meet.[2] In contrast to the Pygmies, these people showed no kindness whatsoever toward each other. They defecated on each other's doorsteps, stole from each other whenever the opportunity arose, and threw their children out of their huts when the children were only a few years old. The adults barely exchanged words but would gleefully laugh if an elderly person stumbled. All in all, their community was bereft of moral values.

Turnbull recognized that morality is a shared belief in behaviors that benefit both the individual and the community. But the Ik demonstrated no respect, even for themselves: "The lack of any sense of moral responsibility toward each other, the lack of any sense of belonging to, needing or wanting each other, showed up daily and most clearly in what otherwise would have passed for familial relationships." Turnbull describes a mother who allowed her own infected pus to spill over her child's food. When he pointed out the dangers of such an act, the mother was taken aback—not because she was hurting her child, but because she could not grasp why anyone would care.

Turnbull tells the sad tale of Adupa, a child who refused to surrender her love for her parents. She would bring them food, which they greedily ate, but they would then refuse her shelter in their hut. And if she cried out in hunger, "they laughed that Icien laugh, as if she had made them happy." Eventually, her parents locked her in their hut and disappeared for nine or ten days. When they returned, they "took what was left of her and threw it out, as one does the riper garbage, a good distance away."

In our society, we would consider such parents criminals, deserving of punishment, not pity. After all, people in their right mind could not possibly do such things, especially to their own children; but one could argue that the Ik were not in their right minds, and that the society in which they were trapped was sick. In Chapter 5, I emphasized that morality is a combination of learned beliefs, neurological development, and peer-group consensus. But something else is needed to maintain moral beliefs, and that is social order. The Ik had none. They had been a nomadic group of hunter-gatherers until the larger society in which they lived had stripped them of their lifestyle and livelihood by forcing them to farm infertile soils in a drought-ridden corner of the world. They had no food and were starving to death, they had given up virtually all hope of survival, and so they had abandoned their moral beliefs.

Today, about thirty years later, some 5,000 Ik still exist, and although their situation remains grim, it does not appear as hopeless as Turnbull originally depicted it.[3] They have learned how to farm, and they even show pleasure and excitement when the occasional foreigner journeys into their mountainous domain. However, according to the American missionary Richard Hoffman, who visited the

Ik in 1996, many tribal members are choosing to abandon their villages, "moving about gaunt and hungry in their tattered cloaks" on their own to search for an alternative life.[4]

The Moral Continuum

Virtually every organized society embraces the edict that you shall not kill or harm other members of your community, or steal from them. Without this rule, which is usually enforced by law and punishment, social order would collapse. Morality is usually defined in terms of acceptable behaviors and how individual actions affect other members of a group. But morality varies between different groups, as do the penalties for moral lapses. Four thousand years ago, the king of Babylonia embedded rules in stone, for the entire community to see. If, for example, the wife of a free man was caught having sex with another man, the lovers would both be tied up and drowned. But if the husband allowed his wife to live, then the king would let the adulterer live as well.[5] What an interesting moral dilemma for a man who deeply loved his wife!

Individuals within the same society often have conflicting moral values. For instance, most people don't believe in lying or cheating, but many easily suspend this belief when taking deductions on their income tax. And most people believe that a right to personal privacy should be respected, except when public figures are involved. Or consider the issue of human rights: why do we have one set of laws governing heterosexual behavior and another concerning homosexuality?

> **Defining Morality**
>
> Human morality is composed of four interconnecting principles: a genetic predisposition toward survival, the neural development of the brain, a social imperative toward group cohesion, and a cognitive propensity to make distinctions between right and wrong and good and evil.

These discrepancies have led me to conclude that we all live along a moral continuum between two abstract poles of good and evil. In every situation in which your ethics must help guide you, you will evaluate your own actions in relation to those of others. Your emotional reactions, personal needs, fantasies, and ideals, as well as the needs of the community, will influence even your deepest moral beliefs. Consider the issue of murder. In principle, we might all agree that it is inherently wrong to take the life of another person, but many people in the United States support the death penalty for a crime that is particularly ruthless or cruel.

Even the foods we choose to eat involve moral issues and decisions. Few people, for example, would tolerate the slaughter of an elephant to satisfy an empty stomach, yet most will not object to killing a chicken or a cow. Why is it "wrong" to kill some animals but not others? After all, we now know that many of the animals we eat on Thanksgiving and Christmas are capable of feeling anger, sadness, depression, and affection.[6] The Dalai Lama once said that he could never eat a plate of shrimp because the notion that so many lives had been sacrificed for a single meal was repulsive. He was reflecting the morality of Buddhism, in which all life is considered sacred. But even Buddhists make moral distinctions of life and death based on their definition of life and sacrifice plants for food though not animals. We might think that plants have less consciousness and experience less pain, but how do we truly know what a plant feels? And when we disinfect our silverware, or take a shower with soap, are we not killing off millions of bacteria in the process? Everyone, it seems, draws an arbitrary line in deciding who and what should live, and why.

Our moral continuum appears to be strongly influenced by the degree of connectedness we feel with others; the more connected we feel, the more we act with generosity, compassion, and fairness. Connectedness also has positive effects on our immune system and emotional well-being.[7] By contrast, lack of connection creates more emotional distance, and so we are less likely to feel empathy toward those we do not know. When people feel distant from others, they can more easily treat others with less respect. Connection guides us to manifest our moral ideals.

Social, ethnic, and cultural differences also contribute to a sense of distance. That is why it is easier to act immorally toward those who do not embrace our own beliefs. Even in our own country,

> For centuries the death penalty, often accompanied by barbarous refinements, has been trying to hold crime in check; yet crime persists. Why? Because the instincts that are warring in man are not, as the law claims, constant forces in a state of equilibrium.
>
> —Albert Camus

which is committed to equal rights for all, nearly every minority group—women, children, blacks, Hispanics, Catholics, atheists, and those with physical handicaps—has had to go to court to clarify our standards of morality. Moral beliefs are continually being redefined by societal law. In the past, for instance, some crimes committed by children and the mentally ill were punishable by death. But in 2002, the Supreme Court reversed an earlier ruling that mentally retarded people could be put to death. For such people, the death penalty is now considered "cruel and unusual punishment." In 2005, the Supreme Court, again reversing an earlier decision, abolished the death penalty for juveniles. To justify the killing of even the most heinous criminal requires a resolution of many competing beliefs.

Speeding Along the Moral Highway

Many of our complex moral concepts take years, even decades to instill in the culture through social conditioning and a strong law-and-order mentality. But neurologically, our brains retain instinctual defense strategies that have evolved over millions of years, and thus we are always, to some degree, at war with the more primitive mechanisms in the brain.

Nonetheless, when an ethical ideal becomes law, people are more inclined to modify their personal beliefs accordingly. For example, few people in America today would seriously challenge a woman's right to vote; but less than a century ago, most men, and a significant number of women, did not believe that women had the capacity to make rational political decisions. And even though women's right to vote was granted in 1920, the number of women in political office has been minimal. Not until 1984 did a major political party nominate a woman for vice president.

Still, when selfish concerns arise we often push aside the law-and-order mentality that we learned in adolescence. For example, many people drive faster than the speed limit because they feel impatient or rushed. Some will justify themselves by saying, "Everybody does it," though as we all know, this rationale crumbles when we are caught. But there are exceptions to this rule.

Some years ago, my uncle was running late, trying to get his wife to an appointment with her doctor, so he decided to ignore the speed limit. Unfortunately, a highway patrol car was right behind him.

The patrolman pulled him over and got out of the car. "Where the hell do you think you're going in such a hurry?" he asked in an antagonistic tone.

Rather than being embarrassed or ashamed, as most of us are when facing an angry authority, my uncle, without batting an eye, replied by firmly saying, "Give me your number."

The patrolman was momentarily taken aback. "What did you say?"

"Give me your badge number," my uncle repeated.

"What do you need that for?" the patrolman asked with annoyance. Authority figures, as I will explain later in this chapter, resent having their authority questioned, no matter what the reason may be.

My uncle then explained, "You used the word h-e-l-l in front of my wife. She is a lady and I cannot tolerate such language in her presence. Now give me your number."

A look of rage, then confusion, came over the patrolman's face, followed by a long period of silence. When he finally spoke, he lowered his eyes and quietly asked for my uncle's forgiveness. My uncle gave it, no citation was issued, and the two cars drove away in opposite directions.

What happened? Why didn't the patrolman issue a ticket? Did he feel that his act was more immoral than the offense of speeding, or did he simply back down for fear of being reported? From a strict moralistic position, one could argue that my uncle should have been given a ticket, since he had clearly violated the law, and that he should have submitted to the punishment because he was a member of society. But when we're speeding along the moral continuum, we rapidly weigh conflicting factors in our decision-making process. Does it seem safe? Will I be caught? Should I be setting an example for others? Unfortunately, if we are in a highly emotional state, we often don't think deeply about possible moral consequences.

Both my uncle and the patrolman exhibited strong emotional beliefs about what they considered right and wrong. In such situations, two competing functions are taking place in the brain: higher moral functioning occurs in the frontal lobes, but powerful feelings tend to suppress frontal-lobe activity, which allows the more primitive fight-or-flight responses of the limbic system (the emotional centers of the brain) to dominate. In my uncle's situation, he may have provoked a sense of guilt in the patrolman's mind by implying that cursing is immoral. My uncle took a risk, assuming that cursing was indeed a moral concern for the patrolman. And the patrolman's own sense of morality may have shamed him into suppressing his anger. Thus the sudden shifts between emotion and cognition helped to slow down any emotional response, and this would allow more time for the patrolman to consider the ramifications of the situation.

Acts of forgiveness can also stimulate frontal-lobe circuits that are associated with compassionate beliefs. This, in turn, further reduces activity of the limbic system associated with anger and fear.[8] We are much more likely to mete out a harsh punishment when we are angry than when we feel compassionate or sad. Angry decision makers do not analyze situations carefully or ponder alternatives; they react instinctually and aggressively, with unrealistic optimism and overconfidence in the rightness of their own actions.[9]

My uncle had played a dangerous game by challenging an authority figure; his challenge could have been easily mistaken as hostility. Yet the officer chose to grant a degree of authority or respect to my uncle, and so two immoral acts (speeding and profanity) were countered by two acts of compassion. For a brief moment, the two men called a truce.

My Aunt the Ex-Con

Moral beliefs are never a private matter, because the acts they engender can impinge on other people's rights. Indeed, many immoral acts are probably committed by people who have no awareness of doing something wrong. How many times have you accidentally taken something from a supermarket without paying for it, stepped in front of someone in a line, or found yourself speeding while listening to a favorite piece of music? This may also happen to children when they take something that belongs to someone else. And it once happened to my aunt—the same one who saw the wolf in her backyard.

My aunt needed to do some last-minute shopping for the holidays, and on her way out the door, my uncle kissed her and gave her a warning. "Be extra careful with your purse because this is the time of year that people steal things."

Off she drove to the mall, and luckily she found a parking space near the entrance. It was just a few minutes before closing time. She dashed inside, bought what she needed, and returned to her car with her packages.

As she was driving home, she noticed that the coin container she usually kept in her car was missing. She thought of her husband's words: "People steal things."

Driving on, she thought some more about how bad people can be. Then she noticed a college sticker on her rear window. "Now why would someone put that thing on my window?" she said aloud. That's when she noticed that the interior was the wrong color and realized that she wasn't in her own car. By a rare coincidence, her key had started someone else's automobile.

Rather than laughing at her mistake, my aunt panicked. She felt that she had stolen someone else's property, and she quickly drove back to the shopping center, leaving the car in the closest spot she could find to the place where she had been before. She acted, in fact, like a child who is afraid of being scolded. She sneaked into her own car and sped off as fast as she could. It didn't occur to her to leave a note on the windshield of the other car. When she got home, she was too ashamed to tell her husband about the incident; and she lay awake all night worrying about how the other driver must have felt when she discovered that her car was missing.

It took a few decades for her embarrassment to wear off, at which time I finally heard the story, which made me ponder some ethical questions. For example, was it wrong to return the car to the parking lot without informing an authority? If I were in a similar situation, I'd probably do everything I could to find the owner and explain the mistake. But many adults behave like children and react with shame when they realize they've done something wrong.[10] They panic and run away or hide, or make up some ridiculous explanation for how their hand mysteriously wound up in the cookie jar.

Right and wrong, like good and evil, are arbitrary when the law does not make matters clear, and the brain is often left to its own devices to evaluate the correctness of its decision. For example, if you found a quarter lying on the sidewalk, would you turn it in to the police? Probably not. But what if you found $1,000. As the emotional effect of a discovery or action increases, we can compromise our moral ideals. The brain must evaluate more issues and possibilities before it can decide what to do. But how do you quantify morality? If you are poor, and your child needs lifesaving medication, is it wrong to take the money from a wallet you find, particularly if you know that the owner is wealthy? When you ask these questions of college students, you'll get a wide range of answers and justifications, for ethical choices are never black and white.

Moral beliefs are usually based on complex rules of logic, reason, social consensus, and personal reward, but these variables can be applied only generally. In other words, each moral issue has unique characteristics that the brain must assess individually. The assessment requires considerable cognitive skill. Does a certain action harm or benefit others? What are the long-term consequences of an act? How do you expect others to act, and how would you feel if someone else's actions affected you? These are just a few of the questions involved in establishing our moral beliefs, and each question involves a series of cognitive and emotional assessments.[11] These processes in turn involve sophisticated frontal-lobe activity. This area of the brain, which is responsible for the executive functions of planning, impulse control, and reasoning, develops slowly during the first two decades of life, so it is not surprising that children and adolescents often exercise poor moral judgment. In teenagers, the developing neuronal connections and strong hormonal and environmental influences may make moral decision making particularly volatile.[12] Moral beliefs may be established in adolescence, but the ability to act on them with any degree of consistency can take many years to refine.

The Moral Brain

When you read a newspaper, the various stories may evoke a range of moral judgments and opinions. It turns out that emotions stimulated by moral issues and those stimulated by nonmoral issues are governed by different parts of the brain. In one fMRI study, the left frontal lobe and the left temporal lobes were activated as people were making moral judgments.[13] The particular areas involved were those associated with working memory, willful thinking, regulating emotions, and abstract reasoning. Another fMRI study showed that simple ethical decision making involves similar areas, including parts of the frontal lobe, and that these areas play an important role in controlling negative emotions such as fear and rage.[14] Frontal-lobe activity is crucial in inhibiting impulsive acts.

When the neural circuits involved in moral assessment are injured, one's ability to interact morally with others can be severely impaired.[15] In one well-documented case, a fifty-six-year-old electrical engineer fell and injured both frontal lobes.[16] Before the injury, he had had no history of psychological or neurological problems, but shortly after his apparent recovery he behaved in bizarre ways. He acted aggressively with little provocation, and he displayed no sense of empathy or remorse. His clinicians reported, "On one occasion, he continued to push around a wheelchair-bound patient despite her screams of terror. . . . His 'lack of remorse' was striking; he never expressed any regrets about the nurses he hit. He failed to accept responsibility for his actions, justifying his violent episodes in terms of the failures of others."

If, say, the highway patrolman who stopped my uncle encountered someone like this patient, he would probably arrest the man. When the frontal lobes are damaged, whether by a tumor or lesion, these lobes are no longer able to reason and process emotions. Patients damaged in this way can exhibit either too much emotion or no emotional reaction at all. In either case, the patients can lose the ability to form moral beliefs or act in morally acceptable ways.[17]

Emotions are essential for making moral and ethical decisions.[18] For example, if you don't have strong feelings about insects, you won't think twice about using a can of Raid to get rid of cockroaches in your kitchen. On the other hand, if part of your belief system honors all living creatures, then you'd probably feel deeply conflicted about this. Instinctual feelings of disgust aroused by the roaches, coupled with fears about disease that could affect your family, would clash with feelings of concern for all life. Would you try to capture the roaches alive and let them go free? Compassion, by the way, tempers the emotional reaction of disgust.[19]

Clearly, moral beliefs play an essential role in suppressing destructive impulses, but such beliefs can also be used to hide emotional problems. The issue of vegetarianism is very interesting because, on the surface, it seems to represent a high moral ideal.[20] Not only does a vegetarian diet spare the lives of animals; it can also be physically healthier. Such beliefs make it easier to maintain a vegetarian lifestyle, but other emotional factors may also be influencing this choice. For example, according to a series of recent studies, adolescent vegetarianism (which has become a fad in many high schools and colleges) can signal an underlying eating disorder that reflects problems relating to self-image and self-esteem. In a study of 4,746 adolescents attending public schools in Minnesota, vegetarians were found to be at greater risk of engaging in unhealthy and extreme methods of weight control.[21] In another study, self-reported vegetarian college women were more likely than nonvegetarians to display disordered attitudes and behaviors related to eating.[22] The same holds true for adolescent vegetarians in Turkey.[23] In these examples, moral beliefs are evidently being used to mask unconscious motivations that are highly charged emotionally.

Guilt also plays an important role in establishing moral beliefs. In a recent fMRI study conducted at a medical university in Japan, researchers identified parts of the brain that are involved with moral evaluations.[24] They found that embarrassment evokes a stronger neural reaction than guilt. This is important, because it helps to substantiate studies in social psychology showing that when one has done something morally wrong, guilt, rather than shame, promotes a greater willingness to change one's behavior. Too much embarrassment and shame can lead to inner hostility and aggressive behavior, whereas guilt—which implies that the person can recognize how his or her acts affect someone else—promotes a willingness to feel sorry for the injured individual. According to June Tangney and Jeff Stuewig at George Mason University, "Guilt is the more moral, adaptive emotion."[25]

The ability to empathize with others is essential for establishing moral beliefs, for if we don't understand how another person feels, we have less ability to respond in a kindly manner. In a recent study, college undergraduates were first induced to feel empathy toward another student, who was actually a confederate of the researchers. When the confederate then insulted the undergraduates, those who had first been encouraged to feel empathy were able to inhibit those parts of the prefrontal cortex that are involved in producing angry responses.[26] Dozens of other studies have demonstrated that short-term empathy-inducing experiments can instill long-term empathic attitudes toward homosexuals, people with AIDS, the homeless, and convicts serving life sentences for murder.[27] Even simply observing another person's facial expression is enough to trigger an emotionally empathic response in the brain.[28]

Unfortunately, as we shall see in some of the experiments that follow, it doesn't take much to evoke hostile, prejudicial, and destructive acts in people who would normally be considered moral citizens. In fact, it takes only a few experimental confederates to manipulate the ethical perceptions in subjects' brains.

How Lies Become Justified

In studies that began more than a half century ago, researchers have used confederates—other researchers posing as study participants—to trick normal, healthy adults into altering their reports of what they see. Take, for example, the drawing of the four lines in the illustration below:

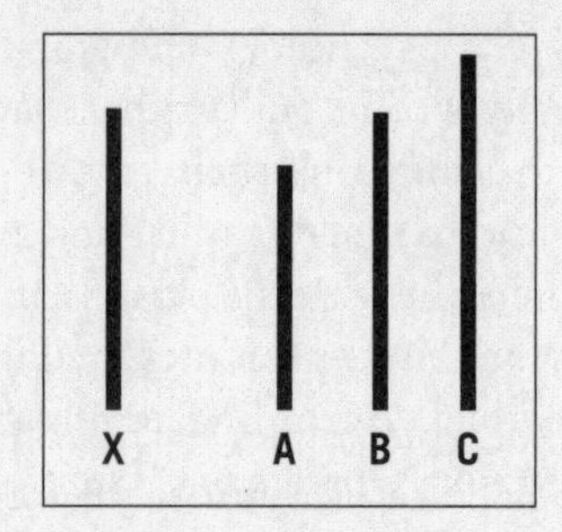

If I asked you to tell me which line is the same length as line X, you'd probably say line B. But what if I were to tell you that you were wrong, that line C is the correct answer? You'd probably experience a moment of doubt, and then you'd go back and double-check. Some people might use a ruler for confirmation before they disagreed with me. But in a larger group of people, something very different occurs.

In a series of trials designed by Solomon Asch, when an individual (the actual test subject) was placed in a room with a group of other people (the confederates) who all agreed that line A or line C was exactly the same length as X, the subject usually went along with the group decision. Of fifty participants, 70 percent conformed at least once, and only 20 percent refused to conform at all. One person conformed eleven out of twelve times. When asked later why they had gone along with group decision, some subjects said they did so in order to please the experimenter. Others had complied because they wanted to fit in. Still others acquiesced in order to avoid social ostracism. But the most interesting finding for me concerned those who genuinely believed that there was something wrong with their eyesight. After all, they reasoned, everyone else couldn't possibly be mistaken.[29]

When there was only one confederate, the conformity rate was only 3 percent. When there were two confederates the conformity rate was 14 percent. With three confederates or more, the conformity rate was 32 percent. Now you can understand what happens in a mob; even people who don't agree will usually participate with the others. Even more unsettling is the fact that when faced with a majority view, many people will not only adapt that view for themselves; they will also convince themselves, and others, of its truth.[30] In one series of experiments, students with normal color vision were shown blue slides, but confederates said the slides were green. As in Asch's study, 32 percent of the subjects went along with the confederates. And in this case even those who refused to comply had a greater propensity to bias their perceptions toward green when they were later shown blue-green slides.[31] These findings, along with those presented in Chapter 5, accentuate the fact that many beliefs, including moral beliefs, can be easily altered by authoritarian and peer-group pressure.

Electrocuting a "Student" at Yale

It's one thing to assume that you can persuade an average person to go along with a mistake or a lie, or even to stand by in silence while a bad deed is being performed; but it's quite another thing to assume that a person with decent morals can be coerced into injuring someone else. In the laboratory, however, it's not that hard to do.

In 1963, Stanley Milgram conducted one of the most controversial experiments in psychology, in an attempt to address a philosophical issue that has haunted politicians and theologians for centuries: why do people obey a command or law that contradicts their fundamental moral beliefs? The experiment, which was initially conducted at Yale, included forty men between the ages of twenty and fifty. First, each participant was introduced to a confederate, "Mr. Wallace," who pretended to be another test subject. The two men were asked to draw lots to decide who would be the "teacher" and who would be the "learner," but the experiment was rigged so that "Mr. Wallace" would always be the learner. The real subject, now designated as the teacher, assisted the research leader by strapping Wallace into a chair and affixing electrodes to his hands. The subject would then go into an adjoining room where he would sit in front of a machine with a row of buttons, labeled from 15 to 450 volts. The higher-voltage buttons were also labeled "DANGER—SEVERE SHOCK." Wallace would be asked questions by the researcher. If he gave wrong answers, the researcher instructed the "teacher" to punish him with increasing electric shocks.

Wallace did not actually receive any shocks, but he responded according to a script. As Dr. Milgram explained:

> At 75 volts, the "learner" grunts. At 120 volts he complains verbally; at 150 he demands to be released from the experiment. His protests continue as the shocks escalate, growing increasingly vehement and emotional. At 285 volts his response can only be described as an agonized scream.[32]

Milgram himself wondered why anyone in his right mind would be willing to participate in giving the apparent shocks, but no one ever declined. Even more astonishing is how far the participant would go in shocking Wallace when prodded by the experimenter. Even when Wallace complained about having a heart condition, two-thirds of the subjects would turn the machine all the way up to 450 volts. Some subjects would literally break out into a sweat as they heard Wallace scream and pound on the wall, but they would continue to increase the voltage, until they eventually heard a loud thump, which was presumably Wallace's body falling to the floor. Silence followed, and as far as the subject knew, Wallace had collapsed or even died.

If the subject questioned the experimenter, or hesitated in applying the shocks, a series of four orders would be given, one at a time, in the following order:

1. Please continue.
2. The experiment requires you to continue.
3. It is essential that you continue.
4. You have no choice, you must continue.

If the participant refused to administer the shocks, the experiment stopped. In the initial study, no one refused below the level of "intense shock," and 65 percent of the subjects obeyed the experimenter and administered "severe" shocks all the way up to the maximum 450 volts. In later experiments, more than 1,000 subjects would be tested, with similar results. But the researchers did discover that if a subject had physical contact with Wallace, obedience would drop to 22 percent. Other studies have supported this finding. The implication is that with increased intimacy (either physical or verbal), people will treat each other with greater compassion and respect.

Milgram's experiment also suggested that moral beliefs, in and of themselves, were not enough to overcome an authoritarian command to commit an immoral act. In the Nuremberg trials, murderers and perpetrators of other atrocities sometimes justified their acts by claiming that they were required to follow orders. But this defense was rejected because there was no record of a single officer being punished for refusing to carry out the execution of a Jew or any other "undesirable." So the threat of punishment might be difficult to use as an excuse.

Reserve Police Battalion 101

In a little town in Poland, in the early hours of July 13, 1942, 500 middle-aged family men were roused from their bunks and taken to a small village where 1,800 Jews resided. Three weeks earlier, these men had been drafted into a special group of "Order Police" by the Nazis. Their commander, Major Wilhelm Trapp, affectionately known as Papa Trapp, informed them, with tears in his eyes, that the battalion had to perform an "unpleasant" task. According to Christopher Browning, a professor of history at Pacific Lutheran University, Trapp told his men that a special job was required of them. Although it was not to his liking, "the orders came from the highest authorities." The Order Police were to take the Jews—women, children, and old people—to a location where they were to be shot. To help assuage the discomfort that the men felt, they were told to think about the bombs that were falling on their own women and children at home.[33] They were also informed that if anyone wanted to decline, he could, but only a dozen men turned in their rifles.

Some records indicate that, at first, only half of the men fired on the prisoners. But with each new assignment, more men would join in. By the end of four months, 90 percent of the men participated. All in all, they shot to death 38,000 Jews. Some even had their photographs taken next to the bodies.

A Century of Genocides

From 1914 to 1923, as many as 2 million Armenians and Greeks were killed by Turks.

From 1934 to 1953, 15 million Russians perished in the gulags.

From 1939 to 1945, several million Polish Christians died in German camps.

From 1942 to 1945, millions of Jews were murdered in Nazi concentration camps.

In the 1940s, as many as 5 million Hindus and Muslims died as a result of India's partitioning.

Since the mid-1950s, 400,000 Tibetans have been killed by the Chinese government.

In 1971, 3 million Bangladeshis were killed by the Pakistani army in less than nine months.

From 1975 to 1979, Pol Pot's regime in Cambodia executed nearly 2 million Cambodians.

In 1994, 900,000 Hutu and Tutsi were killed in 100 days.

From 1991 to 2000, more than 200,000 Bosnians and Serbs were killed in religious wars.

In 2002, the United States accused Sudan of murdering more than 2 million civilians since 1983.

Ordinary men became mass murderers in a matter of weeks. How was this possible? Most people cannot imagine taking part in genocide, but as the box shows, it happened often during the twentieth century. Most people cannot imagine hurting a child, but in the United States alone, nearly 1 million substantiated cases of childhood abuse and neglect were reported in 2001.[34]

The "Prisoners" of Stanford University

We like to think that we are different from people who commit immoral acts, but research indicates that we all are inclined to follow orders from authorities. When questioned, most people will say that if they were in control, they would never act in abominable ways. This popular assumption was shattered in a nearly disastrous experiment conducted at Stanford University in 1971 when the psychologist Philip Zimbardo demonstrated that intelligent people, placed in positions of authority, do quickly abandon their moral ideals.

Professor Zimbardo's goal was simple: to show what happens when you put "good" people into an "evil" situation. First he gathered together twenty-four middle-class men, with no psychological problems or physical disabilities and no history of criminality or drug abuse. He divided them into two groups: "prisoners" and "guards." Each participant was to be paid $15 per day during the two-week experiment, to be conducted in the basement of the psychology department. The basement had been converted into a mock prison, with barred rooms, a "yard" for exercise, and the "Hole," a small cubicle for solitary confinement.

The prisoners' names were replaced by numbers, and the guards were to be addressed as Warden, Mr. Correctional Officer, and so on. The guards were given uniforms and were told to wear silver-reflective sunglasses, which would preserve anonymity and enhance the image of authority.*

In the first stage of the experiment, the prisoners, without warning, were "arrested" in a realistic manner by members of the Palo Alto police department who had agreed to participate in the study. The prisoners were formally booked, then blindfolded and taken to the "Stanford County Jail" (actually the converted basement). There, the newly appointed guards stripped them, searched them, and "deloused" them with a spray. They covered the prisoners' hair with stocking caps (to symbolize a shaved head), padlocked a chain to one ankle, and marched them to their respective cells.

Beyond these initial formalities, the guards were free to create their own rules to maintain "law and order." For example, when a whistle rudely woke everyone up the next morning, some prisoners complained, so the guards decided to make them do push-ups. One guard even put his foot on a prisoner's back. Almost immediately, a protest riot broke out. The prisoners ripped off their stocking caps and the numbers on their uniforms. In response, the guards grabbed a real fire extinguisher and sprayed them with a painful dose of carbon dioxide. Then they stripped the ringleaders and put them in solitary confinement. At first, the "good" prisoners were given special

* A complete description, with photographs and film clips, can be viewed at Professor Zimbardo's website, www.prisonexp.org.

privileges, but by the end of the day, order had yet to be restored. So the guards tossed the "good" prisoners into "bad" cells, expecting to weaken the inmates' solidarity and confidence. The prisoners were even forced to keep buckets of urine and feces in their cells.

By the end of the second day, one prisoner began to have a nervous breakdown, with fits of screaming and crying, but everyone—including members of the research staff—believed that this was merely a trick. It wasn't. Over the next couple of days, as more events escalated out of control, Zimbardo himself began to realize that he was going through a personality transformation: "I began to talk, walk, and act like a rigid institutional authority figure more concerned about the security of 'my prison' than the needs of the young men entrusted to my care."[35] And hardly any other members of the research team questioned the morality of the experiment itself. A Catholic priest who observed the cruelty contacted one participant's parents, who then engaged an attorney to get their son released, but even the attorney was persuaded to let the experiment continue. Finally, the line between role playing and reality became so blurred that the experiment had to be terminated prematurely, on the sixth day. (Today, studies like Milgram's and Zimbardo's would not be allowed, since they can result in substantial psychological harm to the participants.)

In the years that followed his experiment, Dr. Zimbardo attempted to use his findings to improve correctional systems in the United States. But little has changed in the past thirty years. Also, there are disturbing similarities between what happened at Stanford and the abuse of Iraqi prisoners by American soldiers at Abu Ghraib. Put good people into an "evil" place, and morality can quickly break down.

The Seventeen Stages of Evil

The two most significant factors in undermining individual morality are group conformity and the power of authority to override personal objections and doubts. Furthermore, the weaker an individual's moral convictions, the more inclined he or she will be to go along with someone else's beliefs. Drawing from the findings of hundreds of studies on social behavior, the following list highlights the main elements that any person or group can use to sway another person's beliefs and induce behavior that he or she might otherwise refuse to do.[36] Each step after the first builds on the previous one, and the farther down the list you go, the more authoritative, cruel, and ultimately violent group behavior becomes:

1. Establish a set of ideals and beliefs that insinuate your superiority over others.
2. Provide logical justification for implementing your beliefs.
3. Have clearly defined behaviors that the members of your group must endorse.
4. Reinforce steps 1, 2, and 3 as often as possible through discussion and written material until they become your primary beliefs.
5. Have members contractually agree to the above steps—this reinforces a sense of obligation to the group and its leaders.
6. Select a charismatic spokesperson to advertise your group and reinforce your beliefs.
7. Create a range of punishments for those who do not conform.
8. Emphasize the importance of conformity and punishment to help members aspire to your ideals.
9. Insist that each member find new initiates to join the group.
10. Institute severe penalties for those who may wish to leave the group.
11. Limit alternative perspectives and communication between members of your group.
12. Exclude, as much as possible, contact with people from outside the group.
13. Identify a group that opposes your beliefs and ideals.
14. Depersonalize and denigrate those who are not members of your group.
15. Gradually increase hostility and aggression toward the outgroup.
16. When dealing with the "enemy" create a sense of anonymity. (a) Don't use names for your victims. (b) Give impressive ti-

tles to active members of your group. (c) Wear a uniform or a mask, or paint your face.

17. The final solution: eliminate the enemy.

The first five steps can be found in the dynamics of nearly every group, whether social, political, religious, cultural, or societal. Even though there may not be a conscious decision to define one's group as superior to others, this definition happens automatically because of various cognitive functions in the brain. Farther down the list, groups become more authoritarian and cultlike. The final steps are used to coerce members of hostile fringe groups into carrying out socially disruptive acts. However, nearly all these steps are used to various degrees by military and prison authorities throughout the world to carry out their duties and achieve their goals.

With the previous examples in mind, we can conclude that moral behavior is highly dependent on our personal interactions with others. The more anonymous we can be, the easier it is for us to behave immorally. The military knows this well: if you don't see the enemy, it is easier to fire the missile. And if you don't see the bodies, less antiwar sentiment is aroused. Many political scientists and citizens believe that the media's coverage of Vietnam forced the American government to pull out. Later, the military attempted to limit the media's access to events in the Persian Gulf War of 1991, and this issue came to a head again. According to the attorney Kathleen Kirby, of Washington:

> Before and after Operation Desert Storm, the Department of Defense issued regulations governing media coverage of events. . . . [These] made the Persian Gulf regulations the strictest in history. With the conclusion of the first air war against Iraq, the media demanded that the regulations be lifted. The Pentagon eventually responded by imposing a complete news blackout immediately following the start of the ground offensive. The media complained bitterly and filed lawsuits. Subsequently, Washington bureau chiefs and defense officials met to try to arrive at common ground, resulting in the 1992 Pentagon guidelines on coverage of combat operations. The

> guidelines called for, among other things, providing journalists with access to all major military units and to special forces where feasible.[37]

Unfortunately, the key word here is "feasible," since this can be interpreted to suit the moral perspective of those who are in a position of authority. And as I have been emphasizing, authority has a propensity to suppress opposing values and beliefs. Thus the constitutionality of restricting access to war-torn areas has yet to be resolved.

The Trolley Dilemma

A mental game that is often used to show how our moral decision-making processes work is the "trolley dilemma." Here is how it's played. Imagine that you are standing on a street corner, and you witness the following scene. A trolley is running out of control, and in its path are five people who have been tied to the track by a madman. Fortunately, you are standing next to a switch that you can flip, which will take the trolley down a different track. Unfortunately, there is a single person tied to that track. You have thirty seconds to decide: will you flip the switch? If you don't, the five people will be killed. In numerous studies, the vast majority of people say "yes," since they believe that it is worth sacrificing the life of one to save the lives of five.[38]

Now let's change the situation around. This time, you are standing on a bridge, witnessing the same potential tragedy. You can, however, push over the bridge a stranger standing next to you, who will fall onto the track and thus stop the trolley (he's heavier than you, so it will do no good to sacrifice yourself by jumping). The outcome is the same—you'll save five lives by sacrificing one—but very few people would push the stranger. Again, we have the factor of closeness; it's more personal for you to push this man to his death (and therefore morally repulsive), whereas in the first scenario, you are more distant from the victim.

But something more is happening in the brain, for according to fMRI studies conducted by Joshua Greene at Princeton University,[39] subjects take longer to decide when they consider the bridge scenario. The more personal the situation, the longer it takes to make a moral decision; the more impersonal a situation is, the more quickly the brain decides. And for those few people who choose to push the stranger off the bridge, it takes twice as long to make the decision. "According to our model, you've got an emotional response saying, 'no, no, no,' so anyone who's going to say 'yes' will have to fight that response," says Greene, "and you can see it in how it slows people down when they go against emotion."[40] The studies don't tell us whether it is right or wrong to push someone off the bridge, but they do tell us something about how people feel as they struggle to make moral decisions.

When I discussed the "trolley dilemma" with a friend, she said that she could not pull the lever because it would be too disturbing for her to be responsible for the death of anyone, even a single stranger. But when I changed the scenario and told her that her own children were tied to the first rail, she admitted, with sadness, that she'd probably pull the lever. Again, the closer we feel to others, the more likely we are to protect them, regardless of the larger moral picture.

Real-Life Moral Dilemmas

People usually have two responses to each of the following scenarios: a gut-level emotional reaction and a more reasoned answer. The more time you have to contemplate moral issues, the more complex they appear, and the more difficult they are to resolve because the line between right and wrong begins to blur.

Scenario 1

Two men—one rich, the other poor—commit the same crime and are convicted. The penalty is $10,000, or a year in jail. The rich man pays the fine, but the poor man must go to jail. Is this fair?

Scenario 2

You are shopping and notice a stranger slipping an expensive item into his or her pocket. Would *you* report the person? If you see someone in the market eat a handful of grapes and not pay for it, should you report that person? Should people be legally obligated to report a crime, like a petty theft, an armed robbery, a rape, or a murder? Is it morally wrong not to report such crimes?

Scenario 3
You have strong evidence that a captured terrorist knows where a bomb is hidden, and that if it is not found in a matter of hours, it will explode and kill hundreds of innocent victims. Is it acceptable to use physical torture on this prisoner? What measures of inducement should and should not be used? Humiliation? Intimidation? Physical pain? What would constitute cruel or excessive punishment, and how would you measure it?

Scenario 4
A man does not have enough money to pay for his child's medical bill. Would it be morally justifiable for him to lie to the IRS in order to get a refund and thus come up with the money needed to ensure his child's health? Would you ever declare a deduction that was not fully legitimate? If so, how would you justify your act?

Scenario 5
Should wealthy people pay a higher percentage of taxes than the poor, or should everyone pay the same percentage (even though the wealthy can take more deductions and thereby lower their tax)? Should the poor be entitled to more benefits than the rich?

Convincing Ourselves That Something Is "Right"

Our sense of what is fair and right depends on many factors, including bargaining, compromise, and justification. At the Nuremberg trials, for instance, doctors on trial for torturing camp inmates tried to excuse their actions by saying that their subjects had been, in a sense, terminally "ill." Their logic went as follows: because these inmates were scheduled to be put to death, their condition could be considered terminal. And besides, the doctors argued, prisoners could be denied certain rights, such as freedom from inflicted pain.

One outcome of these trials was the establishment of the Nuremberg Code, under which all individuals, whether prisoners or patients, must freely agree to participate in any medical or psychological experiment, and must be adequately informed about the possible consequences. The code expressed, for the first time, the moral belief that science should not be performed purely for the sake of experimentation but should aim to yield benefits to society while not inflicting any unnecessary suffering on experimental subjects.

In 1964, the Declaration of Helsinki prepared by the World Medical Association extended the Nuremberg Code by recognizing that children, the mentally ill, and those who are physically or emotionally impaired should be accorded the same rights and be treated according to the same ethical standards as everyone else.

The primary tenet of the Helsinki Declaration specified that "it is the duty of the physician in medical research to protect the life, health, privacy, and dignity of the human subject." Other important points included the following:

- Everyone has the right to be afforded the best medical treatment available.
- Informed consent is required for unproved treatments.
- The patient has the right to refuse treatment without endangering the patient-doctor relationship.
- If the patient is incompetent, informed consent must be obtained from a legal representative.
- Only volunteers can be used for medical research.
- Research protocols should be established, adhered to, and overseen by an independent committee.

Even with such codes in effect, violations occur, especially when minority groups are involved. For example, the Public Health Service of the United States conducted the Tuskegee syphilis study, which began in 1932 and continued until 1972, ten years after the Declaration of Helsinki had been signed. Four hundred rural black patients had consented to being treated for syphilis but instead of being given what was then the standard treatment, they were given spinal taps, a painful procedure. It was implied to them that these taps were therapeutic. Later, when the deception was revealed, many of the men could no longer be treated, in some cases because the disease had progressed too far. When a class-action suit was filed against the institutions and individuals involved, the researchers tried to argue that syphilis progressed differently in blacks.* The researchers also argued that at the outset, available treatments were not effective. That was not entirely true. According to Allan Brandt, a professor of the history of science at Harvard University:

> The Tuskegee Study revealed more about the pathology of racism than it did about the pathology of syphilis; more about the nature of scientific inquiry than the nature of the disease process. The injustice committed by the experiment went well beyond the facts outlined in the press and the HEW Final Report. The degree of deception and damages have been seriously underestimated. As this history of the study suggests, the notion that science is a value-free discipline must be rejected. The need for greater vigilance in assessing the specific ways in which social values and attitudes affect professional behavior is clearly indicated.[41]

Given the ethical issues raised by the Nuremberg trials, the Declaration of Helsinki, and the Tuskegee study, how should we react to Dr. Klopfer and his patient, Mr. Wright, who was dying of cancer?† The doctor violated a professional protocol by giving Mr. Wright an injection of an experimental drug (Wright was not one of the research patients). Then, when the media reported that the drug was ineffective, Klopfer lied to Wright, telling him that he would be injected with a new, more powerful formula. Klopfer then injected Wright with a saline solution. The patient's life may have been extended several months, thanks to the power of the placebo, but was it immoral to deceive the patient, who eventually suffered crippling

* In a similar manner, politicians have attempted to argue that blacks require different educational standards. The Supreme Court disagreed.
† The complete story was presented at the opening of Chapter 1.

relapses? On the other hand, it seems inhumane to deny a dying man a miraculous cure that he wants.

This brings me back to the question of how people justify their own acts when these acts contradict their moral beliefs. An interesting set of studies found that people will usually act in ways that are primarily selfish, instead of seeking the moral high ground. Subjects were asked to assign two different tasks: one to themselves, and another to a supposed participant who didn't really exist. One task was interesting and offered a reward; the other was boring and would not result in any benefit. Initially, most individuals chose the beneficial task for themselves. Then the researchers added a wrinkle. They told the participants that they could flip a coin to help with the decision-making process. Half used the coin and half did not; but of those who flipped the coin, the majority still assigned themselves the positive task, even when the coin toss went against them. The only thing that changed the outcome was when a mirror was placed in front of those making the assignment. It seems that seeing themselves reinforced the desire to act more fairly.[42]

On a more positive note, another study found that participants who first imagined themselves in the place of the other person were more likely to assign the other person the positive task. Again we see how feelings of empathy and connection elicit greater compassion and care.

Free Will versus Conformity

Moral rules and laws limit our choice of actions, so the question every society, religion, and group faces is to what degree behavior should be controlled. How much freedom should we be allowed? This question assumes that human beings have free will, but it raises the difficult question of where in the brain free will might lie. So far, research has not been able to identify a particular area of the brain that governs free will or self-awareness. Instead, different areas appear to be associated with different types of conscious decision-making processes. Various brain imaging studies suggest that the frontal lobe is critical in directing our ability to act freely and make decisions,[43] and this can be interpreted to mean that free will is a conscious choice—involving an introspective monitoring of the self.[44] This choice is limited mainly to human beings, primates, and some other mammals.

However, other studies suggest that we may have far less conscious choice than we think we have. The ongoing research of Benjamin Libet, for example, has found that several milliseconds before a person makes a conscious decision, there is electrical activity in the brain, which probably represents a subconscious generation of the thought the person is about to have.[45] The implication here is that we do not consciously will things to happen. Instead, our consciousness is more like a video recording of a prior event. Thus, it might be said that we don't exercise free will on a conscious level. To some people, this also means that we cannot be held accountable for our acts.

Even if self-awareness occurs the moment after we act, the rest of the brain will respond to our conscious thoughts the same way that it processes any other incoming stimuli, whether those stimuli are from the world or from other parts of the brain. For example, suppose that someone insults you. In the first few milliseconds, your brain will probably react by formulating a defensive or aggressive response. Before you are even aware of it, you might tighten your jaw and fist, in preparation for a fight. It takes about a half second before your consciousness realizes what you are doing. "Wait! This isn't what I want to do," your consciousness says, as it evaluates a series of physical, emotional, and moral factors. Another second passes as you notice that your fist is beginning to move, but fortunately other parts of the brain are beginning to respond to the messages from the frontal lobe: "No! Stop! Bad choice! Wrong!" A millisecond later, your arm stops moving and the fist relaxes. Meanwhile, in another part of the brain, emotions are still yelling "Defend! Respond!"—so a different part of the brain remembers the saying "An eye for an eye." Before you realize it, you spit out a rude response. Another half-second passes, and your consciousness runs wild: "Oh no, I didn't mean to curse the guy . . . He's bigger than I am . . . I wasn't being nice . . . I should retract what I said." A few milliseconds later, other parts of the brain incorporate this new information, and you suddenly blurt out an apology. Your consciousness, watching everything unfold, congratulates you, "Good going, self!" All this is being encoded into memory so that the next time someone insults you, your brain can access the new memory and thus respond more calmly. Your consciousness has just imprinted a stronger moral code in your brain.

From research on animal behavior, it seems obvious that consciousness (and the frontal-lobe processes that govern working memory, language, and numerous executive functions) is the key to developing moral behaviors and beliefs. But we still have to train the brain to act accordingly, and this can take decades. With enough practice, we may then be able to override the propensity to act immorally when we find ourselves being swayed by destructive influences in society.

Ultimately, the vast majority of people—excluding psychopaths and those with very particular neurological deficits—have the capacity to be aware of how their actions help or harm others. And although each action reflects a combination of conscious and unconscious decision making, we can always influence future moral behavior. From this perspective, we as adults, for the most part, can be held accountable for our immoral acts.

The Psychopathic Brain

Over the years, I have been involved with several criminal cases in which brain scanning was an important part of the defense. In one case, a man was known to have committed a murder. The act itself was not in question; he had admitted it. The issue was whether his brain was so abnormal that it might have predisposed him to commit it. The imaging scans revealed a brain that looked much like what is seen in Alzheimer's disease. The abnormalities were widespread, and they particularly affected the emotional regions of the limbic system.

The defense hoped that these abnormalities would imply that the man was not capable of making adequate moral determinations. Unfortunately, however, other people with similar abnormalities do not commit such crimes or otherwise act immorally. Even if a correlation could be made, you'd have difficulty assessing why one brain-damaged person would be impelled to commit a crime and another would not. Thus most defenses that use this approach will fail.

The same holds true for a defense based on insanity. An abnormal brain is not enough to convince a neuroscientist that the patient cannot distinguish right from wrong. On the other hand, if we could draw connections between neural abnormalities and criminal behavior, we might be able to develop treatments or medications to help certain individuals. Unfortunately, at the present time, the neural underpinnings of psychopathology remain obscure.

In brain-imaging studies of criminals, researchers usually find either a dysfunctional decrease in frontal-lobe activity or an increase in limbic activity. Criminals with decreased activity in the frontal lobes often demonstrate less control over their emotional responses. This decreased control would incline them to act more impulsively in dangerous and illegal ways. In one fMRI study, researchers found disturbed functional activity in the emotional centers of criminal psychopaths who were shown different pictures with positive and negative content.[46] Other studies support the hypothesis that criminals generally have more difficulty using rational thought process to control their emotional responses,[47] whereas other psychopaths may even be able to disassociate from their feelings and thoughts.[48] In one overview of psychiatric disorders, sadistic evil is said to involve a breakdown of frontal-lobe control over the emotional responses arising from the limbic system.[49] Psychopaths whose behavior is strongly antisocial do not become emotionally involved when they consciously and deliberately violate the rights of others.

Recently, HBO broadcast a disturbing documentary that featured a psychiatric interview with an infamous hit man, Richard "The Iceman" Kuklinski, who was linked to dozens of murders and is now serving several consecutive life sentences.[50] As if talking about the weather, he would describe watching rats eat away the faces of his victims. He calmly said that as a child, he would throw pets off the roof because he was mildly "curious" to see what would happen. He claimed that he had consented to the interviews (with Dr. Park Dietz) because he wanted to understand why he acted the way he did; he didn't have a clue. He did have a weakness, he confessed: he loved his wife and children. But for anyone else, he expressed no compassion at all. Once, he shot a stranger through the head with a crossbow just to see if it would work.

One wonders what researchers would find out about Kuklinski's brain if he were to participate in the "trolley dilemma" described above. Studies have found that psychopaths seem to have difficulty processing linguistic information,[51] and that they have more difficulty recognizing the facial expressions of others.[52] In addition, these individuals cannot feel emotions in their body and brain in the same way as most people do. Antonio Damasio calls this the "somatic marker hypothesis." It describes the importance of perceiving our body's response to thoughts, feelings, and behaviors. Some individuals' brains and bodies do not register an empathic response to the suffering of others.

Additional factors that involve immoral behavior are associated with various neurological and psychiatric disorders including seizures, borderline personality disorder, depression, mania, and schizophrenia. For example, one report had to do with two extremely violent and antisocial children; both were found to have small tumors in the limbic system.[53] Once these tumors were removed, the children's behavior improved markedly with virtually no further evidence of violent tendencies. Similar neural disturbances occur from abuse of drugs and alcohol. This strengthens the argument that moral behavior depends on a delicate balance of emotional and cognitive skills.[54]

Evidence from neural research on adolescents shows that brain development is not complete until a person reaches the early twenties, and that the frontal cortex is the last area to mature.* According to the neuropsychiatrist Ruben C. Gur at the University of Pennsylvania, adolescents are "not biologically prepared to exercise mature executive control," and therefore are more prone to act immorally. Gur argues that in a court of law, juveniles "should not be eligible for the most severe punishment available for their crime."[55]

* One index for judging the maturity of the brain is the rate of myelination, a process in which neurons and dendrites are coated with a fatty layer of myelin. This enables neural impulses to travel faster throughout the brain.

Developing Compassionate Beliefs

In summary, moral behaviors depend on wide-ranging networks of interaction, both within the brain and within society. And the degree of interconnection—again, in both the brain and society—will influence the degree to which we exercise our moral beliefs. The value of thinking in terms of a moral continuum is that it allows us to assess each situation in a variety of rational and emotional ways. In other words, each moral issue will be resolved differently, depending on how we are thinking and feeling at the moment.

The more we interact with others in positive ways, the more compassion we will have for them, and the more our moral behavior will be enhanced. This, combined with an increased awareness of morality, will engender a richer perspective that will give us greater control over how we respond to others. In essence, this is what religion and education are all about; and the more we meditate on the social ramifications of our behavior and beliefs, the better prepared we will be to encounter situations that would normally cause us to act selfishly, irrationally, or destructively.

However, once our brain has established our moral system, that system becomes more difficult to modify as we grow older, because during adulthood the brain loses much of its capacity to make new neural connections. This helps explain why the moral fabric of societies changes slowly, and why a society often takes decades to embrace different ideals. As I will often repeat throughout this book, the more you concentrate on a moral ideal, the easier it becomes to act on that belief. Still, the individual will have no way of knowing for certain if a new moral belief will be better than the one it replaced.

Throughout this chapter, I have painted a rather bleak picture of human morality, suggesting that moral beliefs often take a backseat to selfish motivations and acts. We saw tribal parents who treated their children with criminal indignity, students who would lie to please an experimenter, and adults who would cause pain to others simply because they were told to do so. We witnessed caring, intelligent adults transform themselves into cruel authoritarians, willing to humiliate anyone who stood in their way or resisted their commands. And all we have to do is turn on the news to hear about atrocities being committed throughout the world—and of men and women who are willing to murder others because they believe it is the "right" thing to do.

But all these situations took place under conditions of extraordinary stress; and in such circumstances, ordinary people can do extraordinary harm. They'll yell at their spouse and punish their kids because they feel threatened, exhausted, overwhelmed, or out of control. And no one is immune from stress. If we feel pressured by time or money, we're more inclined to break a traffic law, fudge on our tax return, or insult someone we love. Stress not only hurts you and makes you hurt others; it physically atrophies the brain.[56]

Fortunately, the large majority of people uphold moral beliefs and behaviors, and so our civilization continues to work. Considering the challenges we face, human beings exhibit an extraordinary degree of morality and tolerance toward others. Only a small minority cruelly violate human rights. In emotional situations, our moral beliefs may slip, but only temporarily, since most of us learn from our mistakes. We share our struggles with our friends, and we listen as best we can to our enemies, and in the process we unconsciously guide each other toward becoming more socially accepting and at peace.

Part III

Spiritual Beliefs and the Brain

Chapter 7

Nuns, Buddhists, and the Reality of Spiritual Beliefs

ONE NIGHT—A LONG, LONG TIME AGO—AN EXTRAORDINARY event befell a rather ordinary man named Richard Bucke, who had spent a pleasant evening with friends reciting the poetry of Wordsworth, Shelley, and Whitman. On his way home, without any warning, he suddenly found himself engulfed by what appeared to be a "flame-colored cloud." At first, he thought the city had burst into fire, but then he realized that he was experiencing a profoundly powerful inner light.

The year was 1874, and at that time such visual experiences were considered either transcendent moments of enlightenment or evidence of a psychiatric disease. Bucke, who had just turned thirty-six, occasionally had brief but severe bouts of panic, which he diagnosed as "nervous dyspepsia." Here's how he described his panic attacks:

> The first thing the man feels is a great but vague discomfort. Then he notices that his heart is beating much too violently. At the same time shocks or flashes as of electrical discharges, so violent as to be almost painful, pass one after another through his body and limbs. Then in a few minutes he falls into a condition of the most intense fear . . . such that he trembles violently and

> utters low moans; his body is damp with perspiration; his mouth is perfectly dry.[1]

But when Bucke was overtaken by the flame-colored cloud, he felt "a sense of exaltation, of immense joyousness . . . followed by an intellectual illumination quite impossible to describe":

> [I] saw and knew that the cosmos is not dead matter but a living Presence, that the soul of man is immortal, that the universe is so ordered that without any peradventure all things work together for the good of each and all, that the foundation principle of the world is what we call love and that the happiness of every one in the long run is absolutely certain.[2]

The inner light never reappeared, but the event transformed Bucke's vision of the universe and humanity.

Richard Bucke

Today, many skeptics would argue that people who claim to have such experiences are suffering from a neurological or psychological disorder, or are engaged in an elaborate fantasy. But Dr. Bucke's documented biography suggests otherwise. Three years after his experience, he was appointed medical superintendent of the Asylum for the Insane in Ontario, Canada. He became the leading authority in North America on mental diseases. He helped to found a major medical university. Later, he became president of two esteemed medical associations. He wrote and published several books, including a biography of the poet Walt Whitman (whom he resembled)[3] and *Cosmic Consciousness,* which is considered a classic work.

This kindly doctor's experience is similar to many other descriptions of mystical, transcendent, and religious epiphanies; and in this case, there is no evidence to suggest any underlying pathology. Transcendent visions of an ordered, interconnected universe have been known to change people's lives, beliefs, and goals. Bucke, for example, devoted himself to improving the treatment of the mentally ill, and often wrote and lectured on moral development.

Transcendence and the Human Brain

For many years, I have been investigating transcendent experiences like those reported by Bucke. Many religious traditions describe them, as do thousands of intelligent individuals like Bucke. Over the past five years, I have had a chance to examine some of these experiences in our university lab and neurobiologically measure what is happening to the brain as they occur.

At first, I studied Tibetan Buddhist meditation with my late research colleague Eugene d' Aquili. This research provided substantial support for a model of the brain's activity during meditative states.[4] Our findings, which I detailed in my previous book, *Why God Won't Go Away,* were that the altered states of consciousness described by mystics and saints are not necessarily a result of delusional fantasies or the chemical misfiring of a neurologically damaged brain, as many doctors, scientists, and laypeople assume. Instead, these experiences can be brought about when an individual consciously focuses the mind on a sacred image or thought.

These practices, over many years, probably alter the brain's neurological processes in significant ways, and the changes can be recorded in the lab. In this way, we can demonstrate that transcendent, mystical, and spiritual experiences have a real biological component. Furthermore, the neurological changes that occur during meditation disrupt the normal processes of the brain—perceptually, emotionally, and linguistically—in ways that make the experience indescribable, awe-inspiring, unifying, and indelibly real. In fact, the intensity of such experiences often gives the practitioner a sense that a different or higher level of reality exists beyond our everyday perceptions of the world. Many times, such experiences are interpreted within the context of religious beliefs, but many nonreligious people have interpreted them in more secular ways. Bucke, for instance, was a social Darwinist, and in his biography he wrote that even as a child he never "accepted the doctrines of the Christian church." Yet he never doubted the transcendent "truth of what was then presented to his mind."[5]

After Gene d' Aquili and I completed our study of Buddhist meditation, we wondered if other spiritual practices would result in similar changes in the brain. At the time, I was working with several colleagues who were studying the relationship between psychology and religion, and they knew a group of cloistered nuns who practiced a Christian meditation called the "centering prayer," a contemplative method that was first described in a fourteenth-century text, *The Cloud of Unknowing*.[6] According to Friar Thomas Keating, one of three Trappist monks who reintroduced this technique to the Catholic community in the 1970s:

> It brings us into the presence of God and thus fosters the contemplative attitudes of listening and receptivity. It is not contemplation in the strict sense, which in Catholic tradition has always been regarded as a pure gift of the Spirit, but rather it is a preparation for contemplation by reducing the obstacles caused by the hyperactivity of our minds and of our lives.[7]

For our purposes, this meditation was ideal because it was similar to the technique used by the Buddhist practitioners. The Buddhists focus on a sacred image, the nuns focus on a sacred prayer, and both evoke a sense of connectedness with a different level of reality. The nuns described their experience as being in the living presence of God; the Buddhists described theirs as entering a state of absolute awareness of the universe.

Sister Sarah

The first nun I interviewed for the study was Sister Sarah (a pseudonym)—a delightful, charming woman who was then seventy years old. Over the phone, I told her about my research, and I explained how the brain-imaging photos would be taken. Then I discussed how we would inject a radioactive tracer into a vein in her arm while she performed the centering prayer in one of the quieter hospital rooms. I asked her if she thought this would be a problem. "God is everywhere," she replied, "so God is also in your hospital." Then she added, "Praying in your lab won't be any different from doing it in a church."

A Passage from the Fourteenth-Century Mystical Text *The Cloud of Unknowing*

And therefore it is, to pray in the height and the deepness, the length and the breadth of our spirit. And that not in many words, but in a little word of one syllable. And what shall this word be? . . . Let us therefore when we will intentively pray for removing of evil either say, or think, or mean, nought else nor no more words, but this little word "sin." And if we will intentively pray for getting of good, let us cry, either with word or with thought or with desire, nought else nor no more words, but this word "God." . . . Fill thy spirit with the ghostly bemeaning of it without any special beholding to any of His works—whether they be good, better, or best of all.

In order to perform our experiments, I had to address several issues. First, the subjects needed to have a minimum of fifteen years experience in meditation. When I asked Sister Sarah how long she had been performing her prayers, she replied, "Fifty-seven years." "She's qualified." I chuckled. We also have a requirement that no study involving radioactivity can be conducted with pregnant women. Therefore, every female subject of childbearing age is required to have a pregnancy test. I did not look forward to explaining this to a nun, but fortunately I did not have to, since all our subjects were postmenopausal. Nor were any of the nuns taking antidepressants, antianxiety drugs, or allergy medications, any of which would have been a problem, since these can alter the blood flow in the brain.

On the day Sister Sarah came to the university, I shared my hope of showing a link between various spiritual practices and the effects that meditation and prayer could have on the brain. This led to a discussion about the nature of her beliefs, which differed significantly from those of the Buddhist meditators. For Sister Sarah, spirituality and prayer were a gift from God; but for the Buddhists, who do not have a concept of God that in any way resembles Christianity, meditation was a means to connect with the underlying reality of life.

Taking Pictures of God?

Scientific research on the nature of religion is bound to stir public controversy, but the nuns who participated in our study believed that God would be quite pleased that we were taking an interest in prayer. They emphasized that no matter what the findings were, these findings would not shake their faith in God. The Tibetan perspective is different; when the Dalai Lama was asked what he would do if scientific studies invalidated his beliefs, he smiled wryly, saying, "I'd simply change my beliefs!" The Buddha, he explained, emphasized that the way we perceive reality is interpretive and that there is no single definitive truth.[8] I wondered if such differences in beliefs would affect the neurological processes during the act of prayer. The answer, I soon discovered, was a qualified "no."

In our studies, we use an imaging technique called single photon emission computed tomography (SPECT). This method measures blood flow in various parts of the brain. The more blood flow, the more brain activity, and vice versa. First, before the prayer begins, a member of my research team will place a catheter in the nun's arm. Then we attach a long line of plastic tubing to the catheter. I'll stand behind the nun, and as she prays, I will inject a radioactive tracer into the bloodstream. The tracer quickly travels to the brain cells, and within a few minutes the body's metabolic activities break it down, leaving a residue that our SPECT camera can photograph after the prayer has been completed. We then take the subject to another room, where the scanning equipment is kept. The computer images reflect what was happening in the brain during the peak moments of prayer.

We actually do this twice with each person. The first time, I ask the participant to sit quietly, without thinking about anything in particular. This is called a baseline scan, and it shows us what the person's brain is doing in a resting, inactive state. These baseline photos will be compared with those taken immediately following the meditation exercise.

For the centering prayer, the nun begins by focusing her mind on a particular prayer, word, or passage from the Bible. Then she closes her eyes and reflects on the inner meaning and spirit of the text. As Sister Sarah explained, "I open myself to God's presence." Approximately forty-five minutes into the exercise, I inject the tracer through the tube and allow her to continue her prayer for another ten minutes. I then take her to the imaging room for the prayer scans. This step takes another half-hour to complete.

How to Do a Centering Prayer

The centering prayer includes elements that are similar to many meditative practices found in different religions. The following technique has been modified to include practitioners from several spiritual faiths. For the original version, developed for Catholic practitioners, go to www.centeringprayer.com.

First, identify what your spiritual goal or objective is. In the traditional technique for the centering prayer, the person consents to receive the gift of God's presence; a nonreligious person might ask for some inner meaning or truth to be revealed. Then choose a word or phrase that symbolizes this goal, in any way that intuitively feels right. Examples include God, Jesus, Buddha, Allah, Elohim, spirit, love, peace, opening the heart, silence, trust, and favorite passages from sacred texts. If you prefer, you may focus on a spiritual presence, or you can simply focus on your breath. You are giving consent for an inner experience.

Sit comfortably, with your back straight and your eyes closed, keeping your awareness on your symbol or your breath. Do not continuously repeat any word or expression. Instead, be aware of all the thoughts, perceptions, feelings, images, and memories that your contemplation evokes; and if your mind wanders too far away, gently return your awareness to your chosen symbol or breath. During the course of your meditation, the symbol may become vague or disappear. That's OK—simply watch what happens next. You don't need to do anything or make anything happen—just let the experience unfold naturally. After twenty minutes, allow your focus to return to your everyday thoughts and activities.

Contemplating the Divine

Next to the scanner, a computer screen displays brightly colored images of several cross sections of the brain. Reds and yellows signify areas of intense activity; blues and blacks signify parts of the brain where little activity took place. Although they are not as clear in the black-and-white photographs here, you can still make out the differences in neural activity between the nun's resting state and prayer state. In the accompanying photo, the darker areas in the frontal lobe and language center (arrows) show increased activity during prayer.

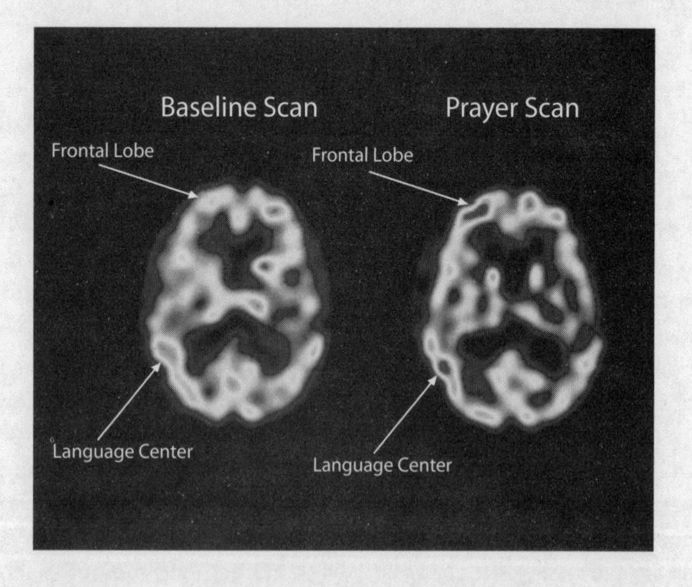

The results were fascinating (the complete study was published in 2003 in the journal *Perceptual and Motor Skills*[9]). Our scans of the

Buddhist practitioners and nuns showed significant similarities and differences in neural processing, with the major difference occurring in the language center of the brain. The nuns had significant increases in activity, most likely because the prayer focused on words and the meanings of these words. There was also greater activity in the right hemisphere, which is involved with the meaning, interpretation, and rhythm of speech (the arrows are actually pointing to the right side of the brain, since brain scans are observed by looking from the feet up).[10] The Buddhist meditators did not show this activity, because they focused on a sacred image, which caused increases in the inferior temporal lobes—the visual processing area of the brain.

The nuns and Buddhists both showed greater activity in the frontal lobes, and in particular in the prefrontal cortex, the part of the brain that is just above the eyes. The frontal lobes monitor our ability to stay attentive and alert, helping us to focus on a task.[11] For this reason, I sometimes like to think of them as the "attention area" of the brain. But they also assist in planning and executing a task, such as reading, running, or meditating on an image or a word. In addition, the frontal lobes play an essential role in processing language, memory, self-reflective consciousness,[12] complex social functions,[13] and pleasure. And, as these and many other studies imply, the frontal lobes play a vital role in the processing of spiritual activities and religious beliefs.[14]

In most forms of meditation and prayer, the practitioner begins with a purpose or goal—to experience God, to calm the mind, to become more aware—which requires increased activity in the attention area. One might say that the act of prayer is a problem-solving device, designed to consciously explore a spiritual perspective or belief and to integrate that awareness into daily life. The attention area would be essential in carrying out such goals, and this was reflected in our scans of both the Buddhists and the nuns. However, something very interesting happens in the parietal lobes that makes experiences of intense meditation unique.

Suspending Time and Space

The parietal lobes, which I often refer to as the "orientation area," interpret sensory information in a way that creates a three-dimensional representation of our surroundings. This gives the body a perception of and an orientation toward where we are in relation to other objects and people.[15] Damage to the parietal lobes causes abnormalities in body image and spatial relations. For example, patients with tumors or lesions in this area may think that one leg no longer belongs to them and have often been found trying to throw this "strange" leg out of the bed. However, if you could consciously decrease activity in your parietal lobes, you would probably feel a brief loss or suspension of self-awareness. You might also experience a loss of your sense of space and time. We discovered that both the nuns and the Buddhists did just that—they were able to deliberately reduce activity in their parietal lobes while meditating. Thus, it should come as no surprise that these individuals describe themselves as entering a state of timelessness and spacelessness, states commonly associated with spiritual, mystical, and transcendent experiences. The following figure shows how the parietal area becomes deactivated (in this illustration, it appears more white during prayer and darker, or more active, during the baseline state) when the nun is at the peak of her prayer experience.

Similar experiences can be triggered when we exercise vigorously

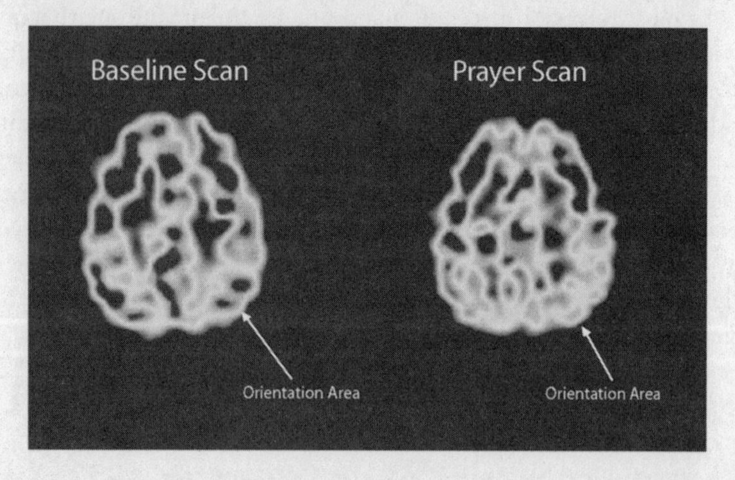

or make love, for at such times we momentarily "lose ourselves" in the experience. We feel more connected to the object of our attention, whether that is a lover, nature, the universe, or God. But how can we compare a rapturous experience of God and the transcendent beauty of a sunset? In essence, we don't, because both can carry significant meaning to the person who experiences them. As I have argued in other writings, spiritual, mystical, and transcendent experiences occur along a continuum ranging from the most subtle to the most profound. They also range from brief feelings of connectedness with something greater than the self to feelings of a complete oneness with all things. The stronger the experience, the more likely the practitioner will feel a sense of connection to a different reality that exists beyond the brain.

Taking Pictures of Beliefs

When I asked the nuns if they wanted to know the results of their scans, I was surprised—they felt no compelling urge to do so. After all, they replied, they did not need scientific evidence to validate their experience—nothing would change their beliefs. They were open to hearing about my findings, nonetheless, so they let me explain. They seemed pleased with my description, but they took the results to confirm that while in prayer they were immersed in the presence of God. I would have worded this differently: while they were in prayer, their sense of God became physiologically real.

Clearly, the nuns had a powerful belief system that accommodated scientific data in a particular way. As far as they were concerned, I was taking pictures of their brain "on God." The Buddhists, by contrast, used the same information to affirm that their practice helped them to reach a level of pure awareness where they could catch a glimpse of an absolute reality. But that reality did not include a notion of God, because God was not a part of their belief system to begin with. As far as they were concerned, I was taking pictures of inner peace.

This is the interesting thing about our frontal lobes: they can allow a dozen people, all of whom have had the same perceptual experience, to interpret it in a dozen different ways. One reviewer despises a movie; another falls in love with it. A gambler finds pleasure in Vegas; a Puritan finds sin. Same physical experience, different beliefs. Even in medicine, two oncologists looking at the same object can come to different conclusions; one sees a harmless mole while the other sees a melanoma and wants to operate immediately.

Something similar happened to my own research after my previous book was published. Skeptics used my findings to conclude that religious experience was nothing more than a neural confabulation within the brain, and religious practitioners cited my work to confirm that human beings are biologically "hardwired for God."[16]

I was intrigued by these disparate interpretations, so I tried to look at the data in a different way. Rather than taking a picture of God, I began to wonder if I was taking pictures of beliefs or, more precisely, of how certain beliefs influence the functioning of the brain. Now, one might intuitively think that different beliefs would affect the brain in different ways, but the brain scans of the Buddhists and nuns showed a remarkable similarity in neural functioning. So I turned the question around. Perhaps it wasn't the specific belief that was influencing the brain. Perhaps the brain was providing a sense of reality for the contents of specific beliefs, thereby validating them. In other words, prayer and meditation might be a way to make our brain experience certain beliefs as real. This brings us to one of the most important structures of the brain: the thalamus.

Perceiving New Realities

The thalamus is a tiny structure, about 1 centimeter in length, that sits on top of the brain stem deep in the center of the brain. Despite its small size, you would basically be a vegetable without it. You'd be alive, but without any semblance of consciousness. The thalamus regulates the flow of incoming sensory information to many parts of the brain, including the higher cognitive processing centers of the frontal cortex.[17] Normally, one would expect to see a simultaneous increase of activity in the thalamus and the parietal lobes when a person is awake, for this is how we orient ourselves to the outside world. However, our studies found that as the thalamus became more active during the act of meditation and prayer, activity in the

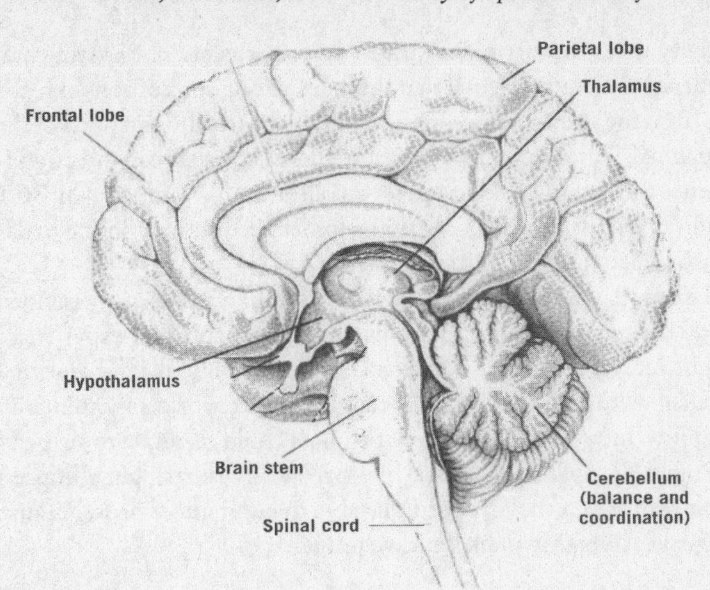

orientation area proportionally decreased. This inverse relationship is very unusual, for even when we are dreaming, both of these areas increase their activity.[18] When we are in a dreamless sleep, or in a state of deep relaxation, their activity usually decreases simultaneously.[19] However, when a person is meditating, ordinary perceptions of the world are being altered, yet the thalamus continues to create a lucid experience. The meditator remains fully conscious, but the brain is experiencing a very different sense of the world.

Here's what happens next: the thalamus communicates this lucid sense of reality to the frontal lobes; the practitioner becomes aware of it and then interprets the experience according to previously held beliefs. Thus the nuns believed that they perceived or experienced the presence of God; the Buddhists felt that they had been in the state of absolute consciousness; and as for nonbelievers—well, some might consider it an anomalous event, a neural quirk or hallucination. Each interpretation is based largely on belief systems the person had developed long before this experience.

For all we know, the thalamus could be responding to incoming stimuli from an unrecognized or unseen source (which some people might call God), but it could also be responding to the conceptual activity that is occurring in various parts of the brain. Experimental evidence with rats suggests that there are strong connections extending from the attention area in the frontal lobe to different parts of the thalamus. This implies that our thoughts and beliefs can directly influence the reality-making processes of the brain.[20] Our own SPECT studies have also found a relationship between activity in the thalamus and activity in the attention area.[21]

I suspect that if a person could maintain a more open-minded state, the range of interpretations concerning spiritual experiences might increase. In many eastern traditions, one will find spiritual teachers who believe that all perceptions of the world are essentially cognitive interpretations. If practitioners could meditate to suspend the brain's propensity to make interpretations, they might glimpse a truer reality. But they wouldn't be able to put it into words, because language is a highly interpretative process.

Creating Emotional Realities

One way our thoughts create a sense of reality is through the regulation of our emotional responses.[22] For example, if we have a pleasant or optimistic thought, this can stimulate a relaxation response, which causes the release of the pleasure chemical dopamine. In that moment, your brain assumes that the world is safe. On the other hand, anxious thoughts send a different message to the emotional centers of the brain, putting the body in a state of alert and releasing various stress hormones—the flight-or-fight response. It doesn't matter whether the anxious thought is based on an actual external threat or a fantasy; the brain assumes that the thought is real and responds to it. In this manner, how we think and the emotions our thoughts stir up deeply influence the way we perceive the world.

This suggests that an anxious person could benefit by engaging in meditative practices that trigger the release of dopamine. In one study,[23] a 65 percent increase in dopamine was found when individuals practiced yoga nidra, a form of meditation in which a person maintains conscious awareness while remaining in a state of complete rest. Other forms of yoga, such as kundalini and tantra (or vigorous practices such as Sufi dancing and chanting) are neurologically stimulating, and thus might not be appropriate for an anxious person. But they might benefit someone who is depressed, since more vigorous forms of meditation and movement can stimulate mood-enhancing hormones and neurotransmitters.

Unfortunately, there is only limited research on a few styles of meditation, so no conclusions can be drawn about what types of meditation might be right for a particular person. Above, I mentioned that an anxious person might not do well with highly vigorous practices, but just the opposite might be true, since I could consider the hypothesis that the energetic practices provide a method for channeling or releasing nervous energy in a positive way. I suspect that each style has a different emotional impact on the brain for different people. But these practices all have one thing in common: each alters our everyday experience of reality.

During meditation, when you first succeed in altering the normal processing of everyday reality, like our advanced practitioners, the emotional centers of your brain are probably going to light up—not enough to cause alarm, but just enough to make you alert. After all, those parts of the brain that monitor reality become very active when things begin to change in unusual ways. Several studies confirm this hypothesis. For example, the results of a study at Harvard using fMRI scans showed increased activity in the regions of the limbic system during meditation, as well as other structures involved in attention (the prefrontal cortex) and the control of the autonomic nervous system.[24]

Two investigators, Saver and Rabin, hypothesized that the ability of the limbic system to label something emotionally as awesome and powerful is a primary driver for why we call something spiritual and why we call something real.[25] When there is a strong emotional response, we pay more attention to it because emotions are perceived as real. For example, let's say you have a vision of a ghost, spirit, or saint. It doesn't matter whether you are dreaming or awake; nor does it matter whether you believe or don't believe in spirits, saints, or ghosts. As far as your perceptual and emotional centers are concerned, the visual experience feels real.

But there is more to our emotional responses than just what is going on in the brain. The autonomic nervous system that connects the brain to the body allows the body to experience both arousing and calming responses. After profound meditative states, practitioners report having felt both intense arousal or ecstasy and deep calm and tranquility. For instance, Richard Bucke's memoirs clearly reflect this, as do the writings of many saints like Teresa of Avila and John of the Cross. In fact, most forms of intense prayer or meditation can make you feel wide-awake and calm at the same time. This is an unusual state of awareness, when contrasted with the body's reactions to everyday thoughts and activities.

A strong emotional response (which is created by the amygdala and other parts of the limbic system) enhances the realness of an event. For example, anxiety will make a minor accident feel like a castastrophe, and depression can turn a minor setback into a hopeless failure. Mania, however, can make some people feel so high and invincible that they might even tell family members that they are God.

> We have only to believe. And the more threatening and irreducible reality appears, the more firmly and desperately we must believe. Then, little by little, we shall see the universal horror unbend, and then smile upon us, and then take us in its more than human arms.
>
> —Pierre Teilhard de Chardin, paleontologist and priest

To summarize, when a meditator focuses on a specific belief or object, the amygdala tells us this is something important, the autonomic nervous system kicks in, and the thalamus makes it all feel real. This information is sent back to the frontal lobes, where it is consciously recognized and then reinterpreted to fit the practitioner's belief. It's also being recorded by the hippocampus, which helps to embed the experience into long-term emotional memory. Of course, this is an oversimplification, but it gives you a sense of how a specific network of circuits can make any experience feel real. Finally, the more a person meditates, the stronger the memory of an event becomes, even when it is recalled many years later. In Richard Bucke's case, the realness of the spectacular night stayed with him for the rest of his life. In fact, any intense experience, if maintained for more than half an hour, can leave permanent changes in the neural circuits involving emotion and memory. If the experience is frightening, the memory can continue to traumatize the individual for years.

The end result of all this brain activity is that the object of contemplation merges with a vivid sense of reality. This suggests that any closely held belief—whether personal, relational, political, spiritual, or scientific—will eventually become a personal truth. This is an attractive scientific hypothesis because it can be tested easily in the lab using the protocols we established for the Buddhists and nuns. However, my hypothesis does not negate the actual realness of any given experience. I have explained only how the brain determines what is real and remembers it. This does not prove that something is or is not real. A brain scan of a nun who experiences God's presence can show only what happens in the brain during that experience. It does not prove or disprove that God exists or that God was actually present in the room. But the brain scan does help us to understand what occurs neurologically when she is contemplating the nature of God.

Born to Believe?

One of the most unusual findings of our scans was that while the nuns were resting (not praying), the activity in their thalami (there are actually two, one on the left and one on the right) was asymmetrical, with one side more active than the other. We found the same anomaly with the Buddhist practitioners. At the University of Pennsylvania Medical Center, my "day job" is to oversee and evaluate scans from thousands of patients admitted to the hospital, and I don't think that I have ever come across a similar asymmetry in a group of people in the ten years I've been here. This asymmetry was so striking that I wanted to review other scientific research on the thalamus. When I did, I could find no similar asymmetries; in fact, the only abnormalities noted in the thalamus involved patients with neurological damage caused by seizures or tumors. In these disorders, the thalamus did not cause the problem; rather, the disease caused disturbances in thalamic activity. Our subjects, however, were normal healthy individuals, neurologically speaking. What, then, might account for this unusual activity?

First, it is important to recognize that each structure in the brain is intricately connected to other parts, and that our consciousness emerges from the neural interactions of the entire brain. Thus, each state of consciousness—dreaming, meditating, or solving a moral problem—is characterized by a different pattern of neural firing. Different rhythms and oscillations in the thalamus, for example, are related to different states of sleep and wakefulness, and any disruption in these patterns can alter the way we perceive and interpret reality and the world.[26] In other words, depression, obsessive-compulsiveness, and diseases related to aging often involve permanent disturbances in thalamic activity.

For example, there is evidence that epileptic seizures alter thalamic rhythms.[27] This may account for the realistic visions that epileptic patients occasionally report. For that reason, many religious visionaries—such as Ellen White, cofounder of the Seventh-Day Adventist Church—have been suspected of having epilepsy. In Mrs. White's case, she did receive, as a child, a traumatic blow to the head that left her unconscious for many days. After this accident occurred, she had many of the symptoms associated with epileptic seizures.[28] As an adult, Mrs. White would sometimes be awakened by religious visions and revelations, and would occasionally make predictions concerning earthquakes, Christ's imminent return, and war. "God is punishing the North, that they have so long suffered the accursed sin of slavery to exist; for in the sight of heaven it is a sin of the darkest dye," she said in 1863. "God is not with the South, and He will punish them dreadfully in the end." But I and other researchers, including many within the Seventh-Day Adventist Church, would not say that such pseudo prophesies (most are vaguely worded, or contain information that can be discerned in nonvisionary ways) are necessarily symptomatic of epilepsy, especially when we consider the religious fervor of the times. Doug Hackleman, who was the editor of *Adventist Currents*, a now defunct magazine of the church, disagrees:

Even though Ellen's trances probably were not the kind of visions she believed them to be, she clearly was a person of vision. She envisioned medical institutions, schools, and publishing houses in various locations around the world; suggested far-reaching changes in denominational organization; and demonstrated at times great insight into the mission of her church. She advocated health care and advanced education for her people. Yet it will be difficult to rightly understand Ellen and what she wrote unless one recognizes the presence of the temporal lobe epilepsy from which she apparently suffered her entire adult life, and that so markedly influenced her thinking, writing, and behavior.[29]

Our subjects showed no signs of any neurological disorder. Instead, the asymmetry we found with the nuns and Buddhists suggests that they have a unique perception of reality, which is continuously active whether they are meditating or not. As one of the nuns commented, "I feel God's presence every minute of the day." Since no long-term studies have been conducted with people who are being trained in meditation, we do not know whether our subjects were born that way—this would imply that they have a biological predisposition toward meditation, religious experience, and perceiving the spiritual realm—or whether intense meditative practices permanently alter the thalamus so as to allow certain states of consciousness to be experienced as real.

As a side note, people who engage in informal meditation—who only attend weekly religious services or yoga training or practice short-term relaxation techniques lasting less than thirty minutes—show no consistent signs of thalamic asymmetry. So the question remains: to what degree can meditation or prayer alter our fundamental perceptions of the world? Science, as yet, cannot say.

A Reality-Making Process in the Brain

Most of the beliefs we have talked about in previous chapters reflect a bottom-up processing of information. Sensory information comes in through the body and is channeled through dozens of perceptual and cognitive processes that analyze, dissect, and reconstruct the information into an internal reality which allows us to function in the world. Much of this information processing takes place in the parietal, occipital, and temporal lobes, which are primarily devoted to preconscious perceptions and long-term memory.[30] But something different happens when this information reaches our frontal lobes. Here we construct a version of reality that does not need to be directly associated with incoming perceptual information about the external world.

It is this version of reality from which our conscious beliefs emerge. In fact, there is growing evidence that the majority of our conscious experiences and beliefs are based on frontal-lobe processes. For instance, there is evidence from fMRI brain scans that the prefrontal cortex is the central repository for working memory, an odd little function that brings a momentary spark of consciousness to a few memories and facts.[31] Once this spark of consciousness emerges, a cascade of neural activity takes place, and a top-down experience kicks in.

This top-down process helps to explain how the power of prayer—or, for that matter, the power of any deeply held belief—can influence our overall perception of reality. To show you how this works, I want to ask you to engage in a little experiment by focusing on the following sentence, repeating it fifty times, silently or aloud, for the next two minutes: "The world is filled with loving people." When you are finished, notice how you feel.

> The world is filled with loving people.
> The world is filled with loving people.
> The world is filled with loving people.
> The world is filled with loving people.
> The world is filled with loving people.

Most people will experience a subtle shift in mood, which can be recorded and verified in the lab. Now I want you to do another experiment. This time, take twenty seconds—no more—to focus on the following negative thought:

The world is filled with mean, selfish, arrogant, violent people.

What do you feel now? You probably lost the earlier fleeting sense of well-being. Instead, you should have noticed a mild sense of irritation. Brain-scan studies find that it takes less than 1 second for a word or a phrase to trigger an emotional reaction in your brain; but I didn't want you to spend several minutes focusing on something negative, because no one knows how long it takes to calm down after experiencing a negative emotional state. Negative states stimulate intensive limbic activity, and this causes the hippocampus to embed the experience into long-term emotional memory. Pleasant experiences, however, do not trigger as strong a reaction. That is why they are harder to remember than unpleasant ones.

These findings suggest that if you want to maintain a sense of well-being, you have to work at it by continually reinforcing positive feelings and beliefs; and this is one of the benefits provided in religious rituals. When you meditate or pray, here's what's happening in your brain: incoming sensory information from the outside world is tuned out; you become oriented solely toward the positive feeling and thought; time, space, and the sense of self begin to blur and fade; a release of dopamine increases your sense of well-being; and this stimulates additional positive thoughts. Voilà! A new sense of reality—i.e., truth—awakens in your frontal lobes, reinforcing the strength of your original beliefs.

A different neural response occurs when you focus on a negative, depressing, anger-provoking, or fearsome thought. Time and space dissolve, but your sense of self is not lost. Rather than feeling connected to the world, you feel independent, isolated, or alone. You may feel in control, or out of control, but in either case, stress hormones and neurotransmitters are being released that will stimulate the defense mechanisms in the brain. It all feels very real, and this sense of reality reinforces the original negative belief.

Of course, our positive or negative beliefs do not affect the reality outside the brain; but these beliefs certainly do affect how we perceive reality. From the research gathered so far, I would say that it takes far more work to generate a positive experience than a negative one. To make matters worse, if you stay in a negative state for an extended period of time, the stress chemicals will physically damage and atrophy different parts of your brain, such as the hippocampus.[32] Over time, you might even lose your ability to return to a calm, peaceful state.* Then your emotions, not your logic and reasoning, will predominate.

A rousing sermon, a political rally, a musical performance, or a very dramatic movie can also stimulate positive and negative responses in the brain. These altered states of consciousness do not last long—fortunately. If they did, you would have difficulty functioning in daily life. For example, it's not practical to be immersed in the loving presence of God, or enraptured by the beauty of a spectacular sunset, while navigating through rush-hour traffic. In such situations, it is essential to maintain a sense of self in relationship to time and space—and, obviously, to the other cars. Nor does it help to get angry. Anger will make you ignore your own bad driving while you condemn everyone else's driving skills.

> When you are stuck in traffic, the great Buddhist teacher Thich Nhat Hanh suggests that every time you see a brake light flashing, you use this as a reminder to take a deep breath and relax. Neurologically, this disengages the emotional centers of the brain so that you can navigate calmly and strategically.

From what we know about the nature of human consciousness, long-term unity of beliefs and attitudes is highly dependent on consistency of frontal-lobe activity.[33] But when something goes wrong, our beliefs—along with our capacity to think logically and morally—become fragmented, and we lose control over our impulses. Anger is perhaps the most dangerous emotion we have because it forces us to think, unknowingly, in narrow, superficial ways. When

*Some evidence suggests that antidepressant drugs can prevent and even reverse hippocampal damage from stress.

we are angry, we do not communicate with much clarity or depth of understanding, yet we feel self-righteous and justified in maintaining our negative beliefs.[34] However, many studies have found that gentle forms of meditation, yoga, and prayer can interrupt destructive emotions and thereby reduce stress.

Transforming Beliefs into Actions

If you believe in a compassionate, loving God, then focusing on this belief should trigger a pleasant, peaceful state. If, however, your image of God is menacing and vengeful, meditating on that belief would evoke a neurophysiological reaction of anxiety and fear. On the other hand, if you become convinced that "God is on our side," this can generate enough anger toward your enemy to initiate a fight. Thus, depending on how you choose to meditate or pray, you can foster compassion or hate; but the key to creating any reality is based on a concentrated repetition of ideas. This does not necessarily require intense meditation, but many types of rituals can evoke very strong responses.

If you want to make terrorists, the formula is unfortunately very simple. Take children or young adults and isolate them from their family and friends. Teach them that their country or cause is great, that they are superior to others, and that the "enemy" is dedicated to tearing them down. You can even introduce the idea of a vengeful God, one who will reward an act of violence against an unholy enemy. Repeat this meditation several hours a day, week after week, month after month. After a few years, these ideas will feel utterly true, and the belief will take on a reality of its own. With body and brain in a state of constant alertness and rage, the conditioned terrorists will find it easy, even desirable, to pull a trigger or detonate a bomb—especially if the promised heavenly rewards are great. As the sociology professor Mark Juergensmeyer bluntly puts it, violence can empower religion.[35]

Though to a far lesser extent, advertisers and politicians use a similar formula to create a specific version of reality in the minds of their audience. Repeat a disturbing story about a candidate enough times, and even though it is false, more and more people will come to believe it. And if you can embed a powerful image or slogan in people's minds—through constant repetition and the promise of a positive effect—they may remain loyal to a product for life. My favorite example is from Coca-Cola: "It's the real thing!" Add to this a century of familiarity (the red-and-white can; or earlier, the distinctive hourglass shape of the bottle), and you can turn a brand into an international icon. Think about it: the last time you needed a facial tissue, did you ask for a Kleenex, a Scottie, or a Puffs? The answer isn't surprising, because we've been meditating on Coke and Kleenex since we were kids.

> All we have to believe with is our senses, the tools we use to perceive the world: our sight, our touch, our memory. If they lie to us, then nothing can be trusted. And even if we do not believe, then still we cannot travel in any other way than the road our senses show us; and we must walk that road to the end.
>
> —from the novel *American Gods* by Neil Gaiman

Our awareness measures reality by the degree of neural activity that is occurring in our brain, and the more we stay focused on our object of contemplation, the more real the thought becomes. Ruminate on your favorite dessert for a couple of seconds, and you will begin to salivate. The same holds true for emotions, for the more you obsess on a particular feeling, the more real it will appear to be.

The moral of the story is this: be careful about what you pray for, meditate on, or obsess about, because it may eventually become your personal truth. If you want to make spirituality a central part of your life—if you want to bring peace or compassion or human rights into reality—then by all means focus on these ideals as often as you can. But if quantum theory or psychoanalysis is your cup of tea, then reading, studying, and contemplating those subjects will help transform them into fundamental truths. Science, psychology, and religion all have intrinsic value and personal meaning; and each points us into deeper layers of a reality that we can never fully comprehend, because of the limitations of the brain.

Chapter 8

Speaking in Tongues

> "And they were all filled with the Holy Ghost, and began to speak with other tongues, as the Spirit gave them utterance."
>
> —Acts 2:4

ON NEW YEAR'S DAY 1901 IN TOPEKA, KANSAS, A YOUNG woman named Agnes Ozman, like Dorothy in *The Wizard of Oz*, was about to be transported to a strange and wondrous place—not by a tornado, but through a born-again experience. She asked her teacher, Charles Parham, to lay his hands on her and pray, and when he did, she began to speak in a language no one had ever heard before. Some of the Bible students thought she was babbling, and others thought she was speaking Chinese, but they all agreed that she had been touched by the Holy Spirit and given the gift of "speaking in tongues." On that day was born the Pentecostal movement, which would transform Christianity throughout the world.[1]

Within a few months, news of Agnes's supposed gift from God had spread across America, and thousands of others began to enter similar rapturous states. Some would sing; others would shout and fall to the ground, believing they were exorcising demons within. Here is how Frank Bartleman described a typical day in 1905 at the Azusa Street Mission, a dilapidated building in downtown Los Angeles that was run by a black Holiness minister named William Seymour:

> The Spirit dropped the "heavenly chorus" into my soul. I found myself suddenly joining the rest who had received this supernatural gift. It was a spontaneous manifestation and rapture no earthly tongue can describe. . . . A dozen might be on their feet at one time, trembling under the mighty power of God. . . . Suddenly the Spirit would fall upon the congregation. God Himself would give the altar call. Men would fall all over the house, like the slain in battle, or rush for the altar en masse, to seek God. . . . The presence of the Lord was so real.[2]

Together, Seymour, Parham, and a few charismatic converts stirred up tornados of revivalism that whipped across the country. Enhanced by the power of gospel music, this evangelical revival quickly gained support among many disenfranchised people of the world.[3] But some other Christians saw the movement as speaking the devil's tongue, and today the controversy still rages. Although the Pentecostal movement rapidly diverged into competing theological groups, speaking in tongues is still considered a meaningful sign that a person has given himself or herself to God.

> Breathing strange utterances and mouthing a creed which it would seem no sane mortal could understand, the newest religious sect has started in Los Angeles. Meetings are held in a tumble-down shack on Azusa Street.
>
> —*Los Angeles Times*, April 18, 1906

In March 2005, my laboratory was given the opportunity to conduct the world's first brain scans on people who were in the act of speaking in tongues. I was approached by a production crew from *National Geographic* who were creating a documentary about people who claimed to have been possessed by evil spirits.

At first, I did not want to participate, as my work focused on the positive aspects of religious practice, not on demonology. But in spite of my reluctance to get involved with anything even remotely related to states of possession, *National Geographic* persisted. As I thought about it, I remembered a conversation I once had with some colleagues about the religious revival movements that had sprung up during the Great Depression. It suddenly occurred to me that speaking in tongues might be considered a positive form of possession, because the individual believes he or she is taken over by the Holy Spirit. *National Geographic* took a few days to track down a proficient practitioner who was willing to be scanned, and then filmed her in our lab.* Before I describe our neurological results, it is important to distinguish between different groups of Pentecostal practitioners, and different styles of "tongues."

200 Hundred Million Converts?

When I first heard that there were more than 200 million Pentecostal Christians in the world,[4] I was surprised, for I personally had never met anyone who confided to me that he or she had spoken in tongues. In a brief search on the Internet I found that evangelical organizations presented the highest numbers, whereas academic research reported that far fewer people actually spoke in tongues.[5] Unfortunately, the statistics gathered by evangelical groups rarely distinguish between Pentecostal and non-Pentecostal traditions; nor do they subtract from their totals defunct churches, members who have become inactive or dropped out, or people who later rejected Pentecostal beliefs. Thus the statistics reflect only the sum total of people who have joined, and these numbers are based on information supplied by individual congregations.

So no one really knows what the actual membership is. However, a recent poll conducted by the conservative Barna Research Group concluded that only 2 percent of the American population currently belong to Pentecostal groups.[6] And of those people, only half have actually spoken in tongues.

* The program, *Exorcism,* was shown on the National Geographic Channel in 2005 as part of the series "Is It Real?" We have subsequently presented these data along with that from four other practitioners at the 2006 Annual Meeting of the Society of Nuclear Medicine.

Different Forms of Glossolalia

References to speaking in tongues—or glossolalia, as it is technically called—can be found in the Old and New Testaments; but until the twentieth century only brief references were made. One exception can be found in the tenets of the Mormon Church, whose founder, Joseph Smith wrote, "We believe in the gift of tongues, prophecy, revelation, visions, healing, interpretation of tongues, and so forth." Smith himself said that he spoke in tongues, as did other founders of the church:

> About the 8th of November [1832] I received a visit from Elders Joseph Young, Brigham Young, and Heber C. Kimball of Mendon, Monroe County, New York. They spent four or five days at Kirtland, during which we had many interesting moments. At one of our interviews Brother Brigham Young and John P. Greene spoke in tongues, which was the first time I had heard this gift among the brethren; others also spoke, and I received the gift myself.[7]

Forms of glossolalia have also been reported in the shamanic rituals of many tribal groups throughout the world. In the Pentecostal movement, this is considered the most important sign that a person has fully accepted the apostolic faith of the New Testament. Pentecostals believe that when they become divinely empowered, they will be given the "gifts of the Holy Spirit," which include the ability to prophesy the future and heal others through prayer and touch. According to the tenets of the International Pentecostal Holiness Church, this happens only through "faith on the part of the fully cleansed believer."[8]

What, exactly, is speaking in tongues? What does it sound like, and what does it mean? Even though the practice was widespread in the first half of the twentieth century, very few researchers took a strong interest in it, and not until 1977 did an ethnomusicologist, Jeff Titon, make a recording of this remarkable speech during a Pentecostal revival meeting. The following glossolalic passage, spelled out phonetically, was made from that recording (the phonetic symbol "?" refers to a guttural sound made in the back of the esophagus):

kantášabaravo sántolavo.
ílamašax rábaxo kalarábou.
rišádalabo píta rabása tóyen . . .
šántoraba sátrobaho sárabaho satóya.
ríka sálara sánto labor?siso l?bokolí
risántobo šantyabaDiánte íkolorosi bal?só koloriánti.[9]

In the early years of the apostolic faith missions, their publications often claimed that parishioners spoke in foreign languages that these people had never learned. Researchers call this form of tongue-speaking xenolalia or xenoglossia; but over the years, as linguists disproved such claims, belief in xenolalia died out. Instead, parishioners came to believe that glossolalics were speaking the language of God.

Researchers who have studied glossolalia have not found linguistic evidence that any form of language is being spoken.[10] Rather, the person is loosely stringing together and repeating familiar phonetic sounds. Nevertheless, in some churches, ministers and parishioners claim to be able to interpret the utterances, though in other groups it is only the speaker who privately intuits the meaning. One woman, who belonged to a now defunct church in Los Angeles, believed that she had been directed by God to express her love for the other members. A man believed that the Holy Spirit was guiding him to seek new employment.[11] But most of the time practitioners simply feel that they are being unconsciously guided to do God's work on earth.

In the 1960s, Werner Cohn, of the University of British Columbia, sent a small group of collaborators, including several amateur actors, to observe tongue-speaking at local Pentecostal groups. Dr. Cohn filmed six collaborators as they attempted to speak in tongues while a young man from Trinidad played drums. All six were able to do so, and their description of their experience was similar to depictions by the Pentecostalists, "They did not know what they had been saying or why, [but] they had the feeling that the language was produced through them and despite them."

When Cohn explained his experiment to a Pentecostal minister, and then showed the movie, "the minister felt that this was true glossolalia, but that it came from spiritual sources other than God."[12] Numerous studies, by the way, have shown that rhythmic stimuli can trigger mystical, transcendent, and religious states of consciousness in part by altering the normal activity of the temporal lobes.[13] For instance, drumming and chanting can both help induce mystical states. Since speaking in tongues is often triggered by gospel singing and music, it is reasonable to assume that the altered states being elicited are genuine.

In 1986, at Carleton University, sixty non-Pentecostal subjects were trained to speak "a pseudolanguage" by listening to glossolalic recordings. Seventy percent were able to speak fluently in tongues, and the researchers concluded that glossolalia is a learned behavior, not an involuntary act.[14] They argued that religious glossolalics do not enter altered states of consciousness; but our neurological studies support the hypothesis that religious glossolalia does indeed occur within an altered state of consciousness.

In more recent studies, researchers at the University of London identified two distinct forms of glossolalia.[15] The first is the more dramatic form described above, involving singing, vocal utterances, and ecstatic bodily experiences. This form usually takes place in congregational meetings. However, the researchers found that many practitioners engage in a gentler, almost silent glossolalic prayer that is associated with calm, pleasant emotions. This can be performed informally and frequently throughout the day, while doing mundane activities like driving a car or performing routine chores.

My study, however, would focus on the neurology of the ecstatic form of glossolalia.

Glossolalia and Mental Health

In the first half of the twentieth century, when revival meetings were popular, psychologists thought that glossolalia was related to some form of psychopathology, but the evidence suggested otherwise.[16] People who speak in tongues showed no differences in personality traits when compared with other population groups—no increases in depression, anxiety, mania, or psychosis. In fact, only a small percentage of mentally ill people engage in glossolalia, and when they do, their reported hallucinations often have religious content.[17] And as several recent studies have concluded, there are beneficial psychotherapeutic effects associated with glossolalic experiences.[18]

One study involved nearly 1,000 clergy members of a British evangelical group. The researchers found that the 80 percent who practiced glossolalia had greater emotional stability and less neuroticism.[19] Other studies have not supported the finding that glossolalia has benefits for health, but neither did they find any negative psychological effects.[20] A study of mentally ill patients did not lead the researchers to conclude that glossolalia was symptomatic of any illness, but they did warn that certain people with unstable personalities should avoid such practices.[21]

The subjects we gathered for our study all believed that they received psychological benefits from their experiences. In general, they were gainfully employed, had a strong network of friends, and were active and respected members of the community.

The Pentecostal Brain in Action

From my perspective, there are several interesting aspects of glossolalia that make it intriguing for research. First, it is popular, for there may be as many people who speak in tongues as people who engage in intensive meditation and prayer. Second, it is a highly religious state, and would thus contribute to our understanding of the brain's activity during spiritual experiences. And third, since glossolalia is a form of speech, I could focus the study on how this type of activity affects the language areas of the brain.

I also suspected that glossolalia might mirror, on a neurological level, other mystical and transcendent experiences found in various traditions and cultures. In this context, brain-scan research could shed more light on the interrelationship of belief systems and spiritual experiences. The Buddhist meditators and Franciscan nuns whom we studied conducted their practices in silence, while remaining still. Speaking in tongues, however, involves vocalization and movement, and this suggested that a very different pattern of neural activity would be involved.

When our first subject—I'll call her Sharon—arrived at the lab, we began by discussing what she thought would occur. "Usually it starts while I'm singing gospel music," she explained, "but I cannot deliberately make it happen." Sharon added that once the "tongues" came on, the process could go on indefinitely, or suddenly stop after a brief moment.

We decided to use our basic meditation protocol of taking two brain scans—a baseline recording and another recording during the religious state. Since I really wanted to detect changes that were specific to speaking in tongues, it did not make sense for the first scan to be taken while she was quietly resting. We had done this with the Buddhist practitioners and nuns, since outwardly they remained in a quiet, motionless state during the periods of meditation and prayer. With Sharon, I decided to compare her brain state while she was only singing gospel songs to what took place during the glossolalic state. Both scans would reveal activity involved with religious singing, a degree of dancing, and intense emotions; but the "tongues" state would reveal the neural activity that takes place when an unknown language is spoken and a person is experiencing the "Holy Spirit."

As in our other studies, we placed a small catheter in Sharon's arm and attached a long line of plastic tubing that would allow us to inject the brain-imaging material later. One of our nuclear medicine technologists—I'll call her Julie—asked if she could assist because she, too, had a deep interest in observing someone speak in tongues.

Sharon started by singing one of her favorite gospel tunes. During this rendition she danced and shouted, "Hallelujah! Thank you, Jesus!" After several minutes of her intense singing, we injected her, and she then continued to sing for another fifteen minutes. She stopped, and we took the first series of scans. These took about an hour, and then we returned to the examination room for the second part of the experiment.

I was quite nervous, and not sure exactly what to expect, since I had never seen anyone speak in tongues before. I expressed my trepidation to Julie by asking, "What if nothing happens?" She smiled and said, "Don't worry, it will." But the reassurance didn't help, because if Sharon did begin to speak in tongues, I wasn't sure exactly how to inject her with the radioactive tracer if she started dancing around the room.

Sharon began to sing while swinging her arms and rocking from side to side. How long would it take, I wondered? Ten minutes? Thirty minutes? An hour? Nobody, including Sharon, really knew. In the studies of meditation and prayer, we estimated that it would take about forty-five minutes before the practitioners reached the state we wanted to observe; but in Sharon's case, we didn't have to wait at all. In less than two minutes, Sharon uttered something that was completely incomprehensible. Then she went back to English. Another minute passed before the next utterance, but this time, she was really speaking in tongues. It actually sounded like a foreign language to me.

My nervousness quickly returned, because now I had no idea how long the glossolalia was going to last. With SPECT scan technology, it takes several minutes for the tracer to leave a residue in those parts of the brain that are being activated; and if Sharon's glossolalia stopped before then, we wouldn't be able to get an accurate impression, and the opportunity would be lost. I waited several minutes to see if the experience might stop, and when it didn't, I quickly gave Sharon the injection and left the room.

I looked up, and to my great surprise, I saw that Julie, my technologist, was also getting into the act, singing and moving around. Suddenly, she too began to speak in tongues. "This is incredible," I whispered to another assistant, and we both watched with astonishment as Sharon and Julie continued to speak in tongues for the next fifteen minutes. Now I knew why Julie had wanted to assist.

It was time to take the next series of scans, so we interrupted them, but it took several minutes before we could bring Sharon back to "earth." She appeared exhausted and emotionally drained, though extremely content. When we asked her how she was feeling, she smiled and said, "Blessed!"

I then turned to Julie, who had always seemed to be a quiet, somewhat introverted woman. "I didn't know that you spoke in tongues," I said. She explained that she had been doing it as part of her religious practice for almost ten years. She eventually became one of our subjects.

The Language of the Gods

When I compared the scans of five Pentecostal women speaking in tongues with those of the nuns and Buddhist practitioners, I saw significant differences. Biologically, the most intriguing finding was that when our subjects began to speak in tongues there was a decrease of activity in the frontal lobes—which in the accompanying photograph shows up as less white. The photo also shows increased activity (more white) in the thalamus, compared with the activity when the practitioners were merely singing.

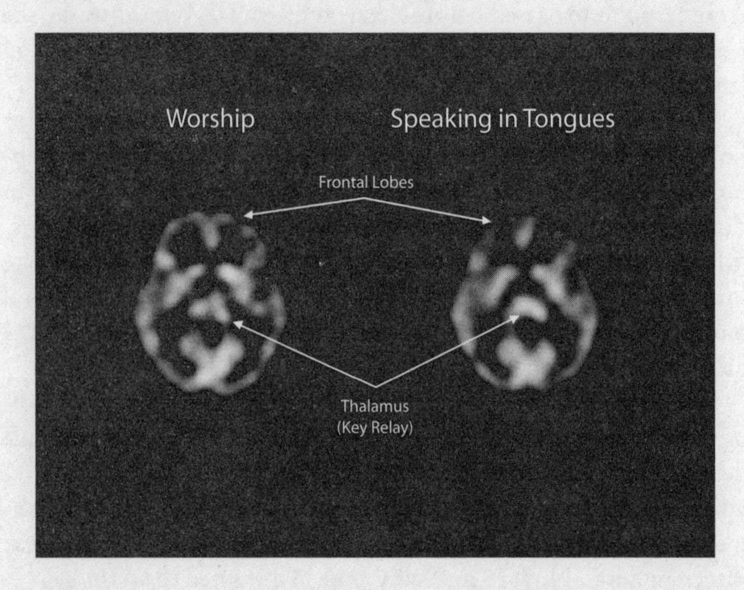

In our studies of meditation and prayer, there was an increase in frontal-lobe function, which I have attributed to the activities used in focusing on specific objects of contemplation. Since the nuns used passages from religious texts, we saw increased activity in the language areas in the frontal lobes; but when the practitioners spoke in tongues, the language areas did not change. One might expect that this would cause disruptions in speech patterns, but the language that emerged during the glossolalic state was highly structured, filled with clearly articulated phrases. This is a very unusual finding, for it suggests that the language was being generated in a different way, or possibly from someplace other than the normal processing centers of speech. For believers, this experience could be taken as proof that another "entity" had actually spoken through them. For disbelievers, it might simply mean that other unique circuits, possibly associated with the thalamus, which directed the style and form of glossolalic speech, were being stimulated.

Language involves complex processes in different parts of the brain, many of which are poorly understood, so one can also argue that the processes involving glossolalia might shed more light on the neurological components of cognition, consciousness, and speech. Another explanation might be that glossolalia is a form of incomplete speech, and thus would not involve the same degree of accuracy that normal speech would require; this too could account for the decreased activity we saw in the frontal lobes.

Suspending Everyday Consciousness

When speaking in tongues, practitioners describe the experience as surrendering themselves to the spirit of God. In this sense, they are no longer attempting to control their thoughts, feelings, or bodily movements; such control is primarily a frontal lobe function. They are also deliberately suspending cognitive processes that are normally active in maintaining focused attention and awareness. In essence, they are surrendering their conscious will. But if their free will is suspended, what then is directing the flow of experience? Again, the believer might say God, but others, like Daniel Wegner of Harvard, would argue that free will is essentially a trick of the mind—a way that consciousness attempts to own or take responsibility for one's behavior and thoughts.[22] Wegner suggests that the involuntary behaviors exhibited during spirit possession, Ouija board spelling, dousing, and hypnosis involve the deliberate suspension of one's sense of will.[23]

With decreased activity in your frontal lobes, you would have the conscious experience that "something else" was running the show. In the study by Cohn mentioned above, even glossolalic people who were trained to speak a pseudo language also had the sense that they were being overtaken by an outside or foreign source. Glossolalia is, in essence, a creative state of mind, and thus allows the person to perceive and interact with the world in a different way. It also frees the individual to interpret his or her relationship with God in nonconformist ways. For this reason, many conservative evangelical groups condemn the practice of speaking in tongues. According to Pastor Bynum of the Tabernacle Baptist Church in Lubbock, Texas, speaking in tongues is never divine and occurs only "where the Bible is not taught and/or understood; where people are seeking experiences, visions and feelings, rather than the truth of the Bible; among people who have been exposed to the cold, dead preaching of modernistic churches; where orthodoxy has become cold and dead; [and] among the lost who are seeking a religious experience."[24]

What Is the Holy Spirit?

According to the Old Testament, the Holy Spirit, sent by God, is the divine principle of activity in the world. In Catholicism, the Holy Spirit (also called the Holy Ghost) is part of the Trinity, a "person" who is both distinct from and a part of God. Other Christian sects believe that the Holy Spirit represents the mysterious and unknowable aspect of God capable of bestowing gifts on individuals, especially the gift of prophecy.

In the practice of fundamentalism, the worshipper, by meditating exclusively on a single religious text, makes that text seem absolutely true. And this reality becomes embedded in the fundamentalist's brain.

Creating New Spiritual Realities

For most religious practitioners, meditation and prayer are designed to reinforce basic tenets and beliefs of the group, and the ongoing activity in the frontal lobes plays an essential role in maintaining and strengthening these beliefs. However, if your goal is to have a new religious experience—e.g., if you wanted God to address a personal problem and directly communicate to you—you'd have to suspend, at least for the moment, your connection to the past. If you didn't, then what you would experience or hear is what you'd expect to experience or hear.

In the Pentecostal tradition, the goal is to be transformed by the experience. Rather than making old beliefs stronger, the individual is opening the mind in order to make new experiences more real. In Sharon's case, first the practitioner interrupts her normal state of awareness by losing herself in the strong rhythmic patterns of gospel music. Memory functions begin to decrease when frontal-lobe activities are suspended; and in this state, the internal rules that govern the construction of language also disengage. But a certain degree of linguistic structure remains. Thus what begins as rambling utterances soon becomes a series of repetitious sounds and cadences that mimic the structure of ordinary speech.

How does the speaker or listener turn such sounds into meaningful messages? In Chapter 3, I described research that found how the brain can turn unfamiliar sounds into words and phrases when it hears ambiguous noise and musical patterns. A similar process could occur in listening to a glossolalic's utterances. Let me give you a simple example of how this works. Read or vocalize the following phrase (taken from Titon's recording of glossolalia) over and over again, until it begins to sound like a chant:

ríka sálara sánto laborʔsiso lʔbokolí
risántobo šantyaba Diánte

Most people, when they hear this phrase repeated, think of the words "saint," "diablo" (the Spanish word for devil), and "dios" (the Spanish word for God). For a person who believes in the gifts of the Holy Spirit, it would be easy to conjecture that the message has something to do with devils, saints, and God—and if the person in question has been suffering, one could interpret the experience as an inner struggle between goodness and evil. Here's another phonetic example of glossolalia, provided by John McGrew of Indiana/Purdue University:

Ke la la iy ya na now.
Key la la iy yey na yey now.

Key la la yey ir now.
Key la la iy ya na key la ya a now.[25]

What language does this sound like to you? To me, it sounds like a Hopi Indian prayer, but in the following example of glossolalia, provided by the meditation teacher George Boyd, the phrases clearly reflect the Hindu tradition in which he was trained (Hindu mantras often conclude with "om shanti," which symbolizes peace):

Umtari amatai, om shanti eem are ah bo ki
ya santai eem keedio em ah la tai, om santie
kam rah to mo ki rai oom santo ray mo ki.[26]

Since messages from God are almost always considered helpful in the Pentecostal tradition, practitioners, when they come out of a trance, may find clarity in the personal issues they have been struggling with. In a sense, speaking in tongues is not unlike the psychoanalytic experience of free association—or dream association, for that matter—because the patient, with help from the therapist, finds meaning in the hidden symbols and disjointed thoughts.

Each person will tend to interpret glossolalic experiences according to his or her primary beliefs about life. The believers' cognitive biases predispose them to interpret the experience in spiritually positive ways, whereas nonbelievers, using similar cognitive biases of the brain, might interpret the experience as a hallucination or a trick of the mind. A social anthropologist, however, might be able to suspend personal bias and see the experience as an integral aspect that maintains cohesiveness in the group.*

When we temporarily suspend our system of beliefs, we may be able to glean new insights about our life that are inaccessible when normal consciousness is governing our thoughts. However, once the rapturous state has ended, normal cognition returns, and this is when individuals would interpret the experience they have had. In this later stage, each person will most likely interpret the experi-

* In Chapter 10, I summarize the different forms of cognitive biases to which human beings are prone.

ence according to the established religious beliefs he or she already holds. Let's say, for example that a woman is grieving over the loss of her husband. Normally, she may feel consciously depressed and lonely, but with the suspension of her thoughts that make up the impression of aloneness, she may instead feel a sense of connectedness that extends beyond her relationship with others. When she comes out of her altered state, she might interpret this "otherness" as God, who reminds her that she is never alone. These thoughts and feelings will reinforce her positive moods, will interrupt negative moods, and may even open the door to new ideas and beliefs.

In a recent overview of the literature, researchers found that the meaning given to glossolalic experiences is closely connected with the social and cultural beliefs of the practitioner's group.[27] In other words, how a person feels and thinks about the experience is greatly influenced by the belief systems of other members of the group. In contrast, for a mentally ill patient who believes he or she is speaking in tongues (or channeling a deity or a demon), the visions and messages are idiosyncratic, disturbing, and bizarre. Usually these experiences contradict the belief systems of the group, and the group will ostracize the person. On the other hand, for someone who has been raised in a shamanic or demon-fearing culture, such possession states may be seen as more socially acceptable.

Maintaining the "Otherness" of God

In our scans of Pentecostal practitioners, we did not find a significant decrease in the parietal lobes, as we had found in the scans of nuns and Buddhists. In fact, activity in this orientation area of the brain slightly increased, which suggests that speakers in tongues do not lose their personal sense of self. The nuns and Buddhist practitioners did lose this sense, and thus they reported feeling at one with the universe or God. The Pentecostal practitioners told me that they never really lose the sense of who they are as individuals; rather, they simply give themselves over to the will of God. They remain in dialogue with God, and thus God retains an aspect of otherness. However, it's possible that if our subjects had continued to speak in tongues for the same length of time as our Buddhist and Catholic subjects meditated, they too might have lost their sense of self.

Maintaining a sense of self is more consistent with Judeo-Christian traditions, whereas Buddhism and Hinduism embrace the notion that the self is an illusion, which interferes with the recognition of our oneness with the universe. This sense of oneness would require decreased activity or a disengagement from the orientation centers of the brain, whereas the activity of speaking in tongues would not require such a decrease—and that is what our brain scans showed.

Making the Holy Spirit Real

Our scans of Pentecostal practitioners also found increased activity in the thalamus, which was also activated in our studies with the Buddhist practitioners and nuns. Because this structure is involved in the transfer of sensory information from the world to different parts of the brain and the body, I have argued that it plays a significant role in making spiritual experiences feel real.

We also found significant increases of activity in the temporal lobes, which contain the limbic areas that monitor our emotions. In other studies, the temporal lobes appear to play an important role in eliciting spiritual experiences. Other researchers have suggested that some individuals have a form of temporal lobe "sensitivity," and that these people are more prone to having visions, hearing voices, and experiencing hallucinations.[28] Furthermore, such individuals are more inclined to believe in the paranormal and the spiritual. In particular, sensed presences are associated with alterations in temporal lobe and hemispheric activity.[29] This may help explain why only half of the members in Pentecostal churches have had glossolalic experiences; the brains of the others may not be built in a way that allows such spiritual experiences.

Since the limbic areas regulate emotions, an increase of activity would be expected when a person was immersed in a glossolalic experience. Gospel singing alone is highly emotional, but speaking in tongues would take the singer to a different level, and add an emotional sense of realness to the experience. Furthermore, it is quite possible that music enhances the reality of the experience. You may have heard of the "Mozart effect." This term is often misunderstood as meaning that music enhances the capacity to learn. Rather, certain types of music, according to various studies, enhance spatial-temporal reasoning and thus give a more holistic sense of space, time, and reality. People think in pictures, not abstractions, and the music helps to bind experiences together within the nerve cells of the brain.[30] Unfortunately, this effect has been scientifically tested only with Mozart's Sonata for Two Pianos in D Major, although it is likely that other musical pieces would have similar effects.

I suspect that gospel singing may first help the practitioner step outside of old memories and experiences. The randomness of glossolalia disorients the practitioner, but the music encourages the brain to have an integrated and spatially realistic experience, as opposed to the dissociative experiences evoked by drugs or psychological distress. In dissociative disorders, the patient feels lost, confused, disoriented, and out of sync with the world, but this is not experienced while speaking in tongues.

Finally, our scans of the Pentecostalists showed increased activity in the midbrain, an area that helps to regulate the autonomic nervous system. Thus it is not surprising that glossolalia, with its vigorous activities of speech and dance, would leave the practitioner in a highly stimulated state, which would be followed by a state of increasing relaxation and calm.

When all the neural circuits mentioned above become activated in ways that are different from our normal processing of reality, we end up with a unique but equally realistic experience of our inner and outer world. However, there is still one element that must be present to make an experience seem spiritually real: we must be open to the possibility that a spiritual realm exists. If we are not, then we will probably interpret the experience as nothing more than a neuropsychological event.

A Different Experience of God

Overall, the experience associated with speaking in tongues seems fundamentally different from the other spiritual experiences we have studied. This suggests that different experiences can lead to different beliefs about religion, spirituality, and God; but it is far more likely that the beliefs we hold about religion and God shape the kinds of spiritual experiences we will have. The Pentecostal movement began with Charles Parham's belief that if the Holy Spirit spoke to and through the apostles of Jesus, his students should also be able to speak in tongues. The more members who believed this, the more likely the experiences would occur. Thus, when Agnes Ozman uttered the first glossolalic words, other members soon began to speak in tongues. The belief triggered the experience, the experience reinforced the belief, and a worldwide movement was born. In other Christian sects, little emphasis is given to this historic religious event, and thus few of their members have glossolalic experiences. Instead, the members have experiences that are closely aligned with the specific tenets of their own church.

All our subjects took great joy in their spiritual practice; and although glossolalia may look chaotic and frenzied to an observer, the subjects all described the internal feeling as amazingly tranquil and peaceful. From a neurophysiological perspective, this is not a surprise, since meditation simultaneously triggers stimulating and quiescent drives. A similar neurological activity takes place when individuals perceive themselves in dangerous situations: they go on alert, yet remain calm. In fact, in a crisis most people feel that time slows down, just as the spiritual practitioners felt in our studies. This happens whenever the brain engages in intense concentration, particularly if a dangerous or exciting situation is perceived. I think that the brain reacts to intense religious experiences in a similar way: if you were expecting to have a potent mystical experience, you'd be excited and intensely alert immediately before and during the experience. Toward the end of the experience and afterward, your body and brain would relax, releasing various calming neurotransmitters and hormones.

Speaking in tongues is a highly arousing way of entering altered states of consciousness. Any religious ritual that involves chanting, dancing, deep breathing, and other vigorous activities could stimulate similar neural activity. Shamans who imbibe powerful hallucinogenic substances while performing healing acts, Sufi dervishes who whirl themselves into ecstatic states of bliss, Nichiren Buddhists who chant to change and create their destiny—all such rituals may engage similar neural processes. However, each practice also stimulates different neural centers, which give rise to different perceptions of reality, different systems of belief, and different notions of reality or God.

Forming Beliefs about God

Neurological studies such as these raise as many questions as they answer, but they also give us insight into how our mind creates and holds on to beliefs. For example, one might ask why a similar neurological experience would cause one person to have one belief about reality and another person an opposing belief. For some people, God represents a force that brings the universe into being. For others, God is a presence that guides human beings to live according to higher moral ideals. Still others see God as the ultimate judge of good and evil acts. There are also those who abandon all belief in God for a naturalistic explanation of the world. How does the brain decide which perception or concept is real?

When different parts of the brain change the level of their activity, they affect the brain's overall perception of reality. If the visual areas are activated, visions can appear (the brain sees a miracle). When the parietal lobe or orientation part of the brain is stimulated, people may have the sense that another presence (God, an angel, or a demon) is near. Decreased activity in the orientation area is associated with an experience of timelessness and spacelessness (God is everywhere) and also a decreased sense of self, so that people feel they are merging with the object of contemplation (being at one with God or the universe). In speaking in tongues, altered activity in the frontal lobes can give a realistic sense that the Holy Spirit is communicating through the self. Finally, increasing activity in the thalamus will enhance that sense of reality, which, depending on the emotional response in the limbic areas, will be perceived as pleasurable, frightening, or exciting. Ultimately, we believe in what we feel is most real.

All this takes place primarily on a preconscious level, but once the information is conveyed to the frontal lobes, a different process begins—one that analyzes, abstracts, categorizes, and quantifies the experience. Other cognitive functions compare the experience with past memories and beliefs, and a reörganization of concepts begins to take place. If the experience is new, then various cognitive processes must evaluate it for future use. Here, in the frontal lobes, the individual consciously decides to accept, reject, or revise beliefs that conflict with the information received. This rearrangement is then fed back into the brain for further analysis and for integration with existing beliefs.

This process takes place in anyone who is actively engaged in religious and spiritual practices. In eastern traditions, individual practice is emphasized, so a person's beliefs are closely aligned with his or her experiences of meditation and prayer. In western traditions, however, participation within the group is emphasized over intensive individual spiritual practice. Usually, such participation is limited to religious group activities, weekly prayer, and the reading of spiritual literature. Nonetheless, these activities can strengthen a believer's inner sense of reality and maintain religious beliefs, whether through personal experience or social conformity.

Unplugging the Neural Connections of the Brain

In the neuropsychology of spiritual, mystical, and transcendent experiences, a particular process, called "deafferentation," often takes place: neural activity in one part of the brain cannot enter another part of the brain. Deafferentation is a complex neural process that has very important functions in the brain. For example, when we focus on a particular task, such as solving a math problem, the brain blocks out incoming information relating to other problems; and if the telephone rings, we can do only one of two things: stop the math and pick up the phone, or ignore it and continue with the math. But if we drop the phone on our foot, our brain will turn its attention to the painful stimuli and not even be aware of anything else that is going on in the room.

Years ago, experiments with isolation tanks showed another dimension of the deafferentation process. In such an experiment, the subject is placed in a darkened soundproof container with no incoming stimulus. When there is no input coming into the brain, it begins to "create" external perceptions from its random functions, thus producing a range of subtle hallucinations. Finally, deafferentation can reflect abnormal functioning caused by a stroke or tumor. Many brain imaging studies have demonstrated that when one area is damaged, changes are observed in the areas that usually receive input from the damaged area.

Different forms of neural deprivation change our perception and analysis of the world, and thus the brain constructs different maps of reality based on the different quantity and quality of the information it receives. This incoherence forces the cognitive centers to interpret the unique experience in different ways, which give rise to different systems of belief. Since there are hundreds of ways in which the neural structures can be deafferented from the input of other functions in the brain, an unlimited number of beliefs about reality can potentially emerge. In essence, our beliefs should be consistent with the reality we perceive, but this is an ideal, because there are always discrepancies between the memories we create and the actual experiences we had.

Transformative Experiences and the Aging Brain

The brain's ability to rewire itself is well documented, but unfortunately the growth that takes place in later life is limited. We all would like to believe that we, like a fine wine, improve with age; but if you've ever sampled a fifty-year-old magnum of champagne, you know that age alone does not improve the quality of the drink. The magnum was served at a friend's wedding, and we all agreed that it was a rare, unique experience—and absolutely horrible to consume. Unlike an old bottle of champagne, we can continue to maintain and refine many of our most cherished skills. We can continue to learn, and we can continue to sharpen our spiritual values. But if we don't work at it, the brain can slip into intellectual and moral stagnation.

The older we get, the less flexible the brain becomes. By the time we reach age fifty, we are less likely to elicit the kinds of peak or transcendent experiences that can occur when we are young. Instead, we are more inclined to have subtle spiritual experiences, and refinements of our basic beliefs. These too, can transform our perceptions of reality in remarkable ways. Abraham Maslow, one of the founders of transpersonal psychology—a field that helped to incorporate spiritual values into the practice of psychotherapy—reflected this view when he compared the emotionally ecstatic experiences of his youth with the "plateau" experiences of his later years:

> For me, part of the loss of peak experiences was the loss of newness and novelty. . . . As these poignant and emotional discharges died down in me, something else happened. . . . A sort of precipitation occurred of what might be called the sedimentation or the fallout from illuminations, insights, and other life experiences that were very important—tragic experiences included. The result has been a kind of unitive consciousness which has certain advantages and certain disadvantages over the peak experiences. I can define this unitive consciousness very simply for me as the simultaneous perception of the sacred and the ordinary, or the miraculous and the ordinary.[31]

Maslow saw the peak experience as a momentary emotional state, "empty of cognitive content," whereas the plateau experience brought with it an ongoing state of serenity. Our study of Pentecostal practitioners speaking in tongues captured, perhaps for the first time, how a person's brain functions during a peak experience. Our studies with the nuns and Buddhist practitioners lend credence to Maslow's plateau experience, in which the individual maintains, through regular practice, a state of serene consciousness that focuses on the sacred dimensions of life.

I imagine that after years of meditative practice and prayer, the mature individual maintains a delicate balance between the cognitive, emotional, and perceptual centers of the brain. Such a person has entertained many beliefs, but his or her absolutism has been softened by years of deliberately suspending it for brief periods of time. Such individuals know that what they see, feel, and think are all partial views of reality, and thus they find it easy to keep an open mind concerning other systems of belief. To me, a living exemplar is the Dalai Lama, because he has often said that if science were to prove his beliefs inaccurate, he would simply change his beliefs. The Dalai Lama is one of the few religious leaders who openly embrace the scientific discoveries of the West, while nurturing relationships with other religious groups. He even talks compassionately about the Chinese leaders who have persecuted his people for nearly fifty years.

In order to keep an open mind, and to feel more connected to others, all you may require neurologically is a conscious reduction of activity in both the frontal and the parietal lobes, along with a suspension of limbic activity relating to anger and fear. So far, the evidence suggests that spiritual practices can be used to either strengthen or suspend old beliefs, but that the process of permanently changing them is far more complex, because so many other brain mechanisms work to maintain the consistency of previously acquired beliefs.

Spiritual Realities Revisited

The question remains: can transformative experiences dramatically change a person's fundamental beliefs? It is intuitively appealing to say "yes," and certainly there are many anecdotal accounts throughout history of nonbelievers suddenly embracing spiritual ideals. But for the most part, those who experience religious epiphanies—like the founders of the Pentecostal movement—have already been believers in God, or at least agnostic. The Dalai Lama, for instance, was considered profoundly curious and open-minded as a child, so one could argue that even he has not fundamentally changed his beliefs.

Tolstoy is a rare example of a person who experienced a religious epiphany and conversion in his later life. At the age of fifty, he wrote, "I did not know what I wanted. I was afraid of life; I was driven to leave it." Still, he had a craving, which he called a "thirst for God," that haunted him for years. And then one day his divided soul was unexpectedly healed:

> I was alone in the forest, lending my ear to its mysterious noises. I listened, and my thought went back to what for these

> three years it was always busy with—the quest for God. But the idea of him, I said, how did I ever come by the idea? And again there arose in me, with this thought, glad aspirations towards life. Everything in me awoke and received a meaning. . . . To acknowledge God and to live are one and the same thing. God is what life is. . . . After this, things cleared up within me and about me better than ever, and the light has never wholly died away. I was saved from suicide.[32]

In a sudden conversion—when a person shifts from disbelief to belief, or shifts from belief to disbelief—the brain itself would also have to undergo a biological transformation in order to accommodate the new outlook. It seems highly unlikely that a researcher would be lucky enough to capture that moment in a laboratory setting, since it is almost impossible to determine when such an event might happen.

Thus the question of how, why, and where transformative experiences occur in the brain—and whether they make permanent changes in neural activity—remains a scientific mystery, with one possible exception. When I went back and asked one of our Pentecostal subjects to submit to a resting-state scan (lying quietly on the table for twenty minutes, without thinking), I again discovered that she had asymmetrical activity between the two sides of her thalamus, very similar to what I found in the resting-state scans of the Buddhist practitioners and nuns.

This, as I emphasized in Chapter 7, is unusual, for it suggests either that the people we have been scanning are born with a unique capacity to have spiritual revelations, or that they have altered their neural functioning in permanent ways as a result of years of intensive practice. And for these people, the experiences they encounter during intensive meditation are as real as anything you and I can touch. For the Buddhist, unitive consciousness is real. For the nun, the presence of God is absolutely real. And for the Pentecostal practitioner who speaks in tongues, the Holy Spirit is not just a metaphor; it is as solid as the book that you are now holding in your hands—as far as the brain can tell.

Chapter 9

The Atheist Who Prayed to God

IN 1988, A YOUNG COLLEGE TEACHER, WHOM I'LL CALL Kevin, had a spontaneous mystical experience that transformed his perception of the world and his career. "I was sitting in my office, which has a marvelous view of the campus gardens," said Kevin, "when I suddenly felt as though a blanket of intense silence had fallen over the scene. Then I had the sense that I was intimately connected to everything—the sky, the trees, the grass, even the garden walls. It was as if everything in the universe was in its proper place. Immediately, I was filled with a sense of peacefulness that I had never known before. I even remember saying to myself, 'So this is what those eastern mystics were talking about.' The experience stayed with me for weeks."

Kevin told me that the experience never happened again. Still, he felt that it changed his life in significant ways. "Before the experience, I had only a passing interest in religion, but since that time, I've been fascinated by the mystical traditions of the world. And even though nearly twenty years have passed since that happened to me, it feels like yesterday when I think about it."

The research literature suggests that most people who have such experiences become more spiritual, but this did not happen to Kevin. Instead, he went from being an agnostic to being a staunch atheist. "All I can tell you is that in that moment, I felt absolutely certain that when I died, that would be the end of it. Nothing. No afterlife, no spiritual dimension, no God. And the feeling left me elated. Later, I rationalized that if I were to accomplish anything significant, I only had this one life to do it in. I couldn't sit back and passively wait for 'it' to happen. If my life was going to change, if I was to become happy, then I'd have to take complete responsibility for bringing it into being. No god was going to intervene."

The experience catapulted Kevin into a decade of research into the nature of religious experiences. He also began to explore a variety of eastern and western meditation techniques, many of which led to other peak experiences and insights, but none of them ever gave him the sense that a spiritual realm existed. "I wish I believed in God—I really do—but my experience doesn't support such a belief. So I tend to see God as a psychological function of the mind, though I must admit that I am rather envious of those who do believe. I think it makes life easier to have that kind of faith."

Kevin responded to an invitation I circulated to several atheist organizations to participate in our research studies. Kevin interested me for three reasons: he was an atheist, he meditated regularly, and he often had experiences that were similar to those described by those who have maintained long-term spiritual practices. He's exactly the sort of person I wanted to interview and scan. But why would I study atheists? After all, atheism is not a religion; it is a thought system that reflects disbelief in theological and God-oriented premises. For me, atheism raises the question whether there are neurological differences between those who believe in God and those who do not.* After all, a strong disbelief can influence a person's thinking and behavior as much as a strong belief, and there should be a way of exploring how such differences affect the brain. Since evidence suggests that religious affiliation modestly improves health, studies on atheism might raise the question whether or not an inherent disbelief in God carries with it any physiological or psy-

* Mathew Alper, in his book, *The God Part of the Brain* (Rogue Press, 2000), articulates this concept very clearly by proposing that religious and nonreligious individuals might fall along a "bell curve" continuum, with highly religious individuals falling on one end of the curve, and atheists falling at the other end. Our brain-scan study with Kevin would help to set the stage for addressing this intriguing premise.

chological risks. These are some of the questions that I hoped our lab could explore.

The Most Disliked Minority in America

We medical researchers are always interested in studying exceptional cases, and atheism is relatively rare in America. In most surveys, the conclusion has been that approximately 80 percent to 95 percent of Americans believe in God or maintain other spiritual beliefs; only 1 percent of the population claim to be atheists.[1] Worldwide surveys have concluded that atheists represent approximately 8 percent of the population, but I suspect that the numbers are higher, because people who hold unpopular beliefs are less likely to participate in such surveys. In some countries, as the table here shows, atheism may run as high as 88 percent.[2] In 2005, Phil Zuckerman of Pitzer University carefully assessed the most recent surveys and concluded that between 500,000,000 and 750,000,000 people worldwide do not believe in God.[3] This raises the question why beliefs about God vary as much as they do.

Percentages of People Who Are Atheists, by Country

East Germany	88
Russia	27
Israel	26
Netherlands	24
Hungary	23
Norway	15
Britain	14
West Germany	12
Italy	5
Ireland	3
United States	1

Note: This survey is from 1991. Other surveys show that Japan, Denmark, and Sweden have high percentages of atheists, while South American countries have the lowest percentages. More recent surveys show much higher percentages for people who "do not believe in God."

When we consider that nearly half of all Americans express moderate to strong disapproval of those who have no faith in God,[4] we can understand the reluctance to admit to atheism on any survey. According to the Pew Forum on Religion and Public Life, "nonbelievers are particularly unpopular among the less educated, more conservative, and older segments of society," but even 37 percent of those with college degrees felt unfavorably toward people who didn't believe in God.[5] As we shall see later, it is actually very difficult to assess the religious or spiritual temperament of a country, since much depends on the setting of the survey, how the questions are asked, and what definitions are used.

No one, to my knowledge, has been able to explain scientifically why the mere mention of belief in God—or its opposite, atheism—evokes such a strong emotional reaction in Americans. Since the neurology of emotional processes has been studied extensively, brain-scan technology might shed some light on the emotional circuitry involved in atheism and belief. But we have to remember that brain science is still at an early stage of development. Identifying an emotion that is tied to a specific thought or belief is extremely difficult. Nonetheless, we do know that any idea or belief that is repeatedly recalled and contemplated becomes strongly embedded in memory, taking on greater and greater nuances of reality and emotional import in the brain. In the United States, it is unusual to open a magazine or turn on the television without encountering a religious debate. Even the Supreme Court, in recent years, has been flooded with cases concerning the Ten Commandments, the Pledge of Allegiance, religious charities, door-to-door proselytizing, Christmas, menorahs, and fundamentalism.[6] Such attention raises emotional issues for believers and disbelievers alike, and these issues in turn leave strong neural imprints in the memory circuits of the brain.

If we view religion as a philosophy, then atheism can be seen as occupying one end of a spectrum of beliefs. But even this oversimplifies the concept of atheism, for there are individuals to whom God holds little meaning, yet who do not disbelieve. In this form of "practical" atheism,[7] life is lived as if there is no God. Then there are those who reject the specific theology of the ruling society; they still hold spiritual beliefs, but not the ones that are sanctioned by the prevailing religious authorities. For example, the seventeenth century philosopher Baruch Spinoza was considered an atheist because he did not make a clear distinction between nature, reality, and God. The early Christians also were considered atheists because they rejected the Roman deities of that era. This form of "classical" atheism can be contrasted to what is known as "dogmatic" atheism, in which a person explicitly rejects God's existence. In "militant" atheism, which was popular in the first half of the twentieth century, and was often associated with communism and socialist politics, individuals attempt to undermine the religious foundations of a particular society. Other shades of atheism and non-religiosity also exist, just as there are different shades of spiritual and theological beliefs; and there are even various groups who consider themselves "spiritual" atheists.

> By God, I mean a being absolutely infinite—that is, a substance consisting in infinite attributes, of which each expresses eternal and infinite essentiality.
>
> —Spinoza (1632–1677)

In America, you'll find freethinkers and secular humanists, as well as rabbis and ministers who happen to be firm disbelievers in God. There are even organized congregations that do not embrace any particular deity or spiritual realm. For example, the Unitarian Universalist Church rejects the notions of heaven, hell, and divine intervention. Its membership—which now exceeds 600,000 and once included Jefferson, Emerson, and Thoreau—embraces a naturalistic worldview and the universal ideals of freedom, justice, and democracy, but without a need for the concept of God.[8] For the last 450 years, their goal has been to practice religious tolerance in "the never ending search for truth." Although Kevin is not a member of any religious group, he says that his values are strongly aligned with Unitarian principles.

Are Surveys of Spirituality a Valid Form of Measurement?

Kevin's parents were Jewish, and although they believed in God, they did not attend religious services regularly. After his bar mitzvah, Kevin joined a Jewish youth organization and was involved with various projects aimed at eliminating anti-Semitism; but by the time he entered college, religion held little interest for him and had little meaning in his life.

In college, Kevin sporadically experimented with psychedelic drugs; but he found the experiences unsettling, so he stopped. Like many people of his generation, he occasionally participated in workshops that introduced him to Zen, yoga, and various eastern movement exercises like tai chi, which superficially exposed him to nonwestern spiritual philosophies. But it wasn't until the afternoon of his unexpected transcendent experience that Kevin seriously involved himself in spiritual disciplines. Today, at age sixty-five, he regularly meditates and uses guided imagery to relax, to cope with physical discomfort and pain, and to develop compassion toward others. "It's worked for everything from sinus headaches to surgery, but I can't seem to do it in the dentist's chair," he said with a laugh. "There, I just use headphones and loud music to distract me from the sound of the drill."

Kevin mainly practices a form of insight meditation[9] in which he sits quietly and observes how his thoughts unfold. "I find it very similar to the process of traditional psychoanalysis, but mostly it helps me to turn off anxious feelings," Kevin said. His current objective, which is similar to the spiritual goals of the nuns and Buddhists in our previous studies, is to develop a deeper sense of inner happiness and peace. For this, he actually does a meditation practice that is very similar to the "centering prayer" I described in Chapter 7.

During our interviews, Kevin did not seem anxious or depressed, and his healthy state of mind was borne out by tests previously administered by a clinical psychologist. As a first step in studying Kevin, I administered a series of questionnaires, developed by various researchers, to identify a person's degree of spirituality or religiosity. These surveys are often used to correlate issues of physical and emotional health.* One of the tests, the Index of Core Spiritual Experience, was developed by Jared Kass, director of the Study Project on Well-Being at Lesley University in Massachusetts.† This index incorporates questions developed by previous researchers

* If you would like to see the types of tests Kevin took, or participate in the University of Pennsylvania Survey of Spiritual Experiences, go to www.neurotheology.net, but please note that two other sites—neurotheology.com, and neurotheology.org—are unrelated to our work.

† You can take this test, and compare your score to Kevin's, by going to http://www.spiritualityhealth.com/newsh/items/selftest/item_234.html.

studying spiritual and mystical experiences and has been correlated with increased life purpose and satisfaction and a decreased frequency of medical symptoms.[10] Kevin's score ("inspirit" 18, "well-being" 4.62) indicated that he had a healthy sense of self-confidence and well-being, and that spirituality was likely to be a strong factor in his life.

Kevin, however, complained that he was uncomfortable with how the term "spirituality" was used in these tests. "To me, spirituality has two meanings. Although I sometimes use it to refer to my meditation practice, I do this more out of habit and convenience when talking to others. I actually don't believe in the existence of any spiritual realm." In Kass's survey, the first three questions are designed to distinguish between religious and nonreligious individuals:

> "How often do you spend time on spiritual or religious practices?"
> "How spiritual or religious do you consider yourself to be?"
> "How often have you felt close to a powerful spiritual force that seemed to lift you outside yourself?"

For an atheist, such questions can elicit paradoxical responses, depending on how he or she defines spirituality. When Kevin thought about his Buddhist meditation, he initially stated that he spent a great deal of time doing spiritual practices, and so, within this context, he considered himself a moderately spiritual person. And, since he had had many transcendent and uplifting experiences in his meditation, he also responded positively to the third question, saying that he felt close to a spiritual dimension in life.

Religious and spiritual surveys often do not clearly define their terms, and this vagueness presents a critical problem when one is evaluating religious data. In addition, some questionnaires include so many examples of what may be considered "spiritual" that the term begins to lose all meaning. For example, some studies have included in their lists various pleasurable and nonreligious experiences such as watching a beautiful sunset, being moved by a piece of music, and even gardening. Furthermore, test subjects are rarely asked to define what they mean by spirituality. When I asked Kevin for his definition, he described it as a belief in a nonphysical, immaterial, supernatural dimension, a notion he explicitly rejects. To his way of thinking, all phenomena can be explained through the naturalistic principles of science.

In 2003, the Higher Education Research Institute at the University of California, Los Angeles, conducted a research program tracking the spiritual growth of students during their college years. The researchers concluded that 80 percent of those surveyed considered themselves spiritual, but these researchers had included in their definition anyone who actively sought creativity, inspiration, or meaning and purpose in life. I would argue that most people, atheists included, would embrace similar goals and ideals. Here is a quotation from the project description:

> Spirituality points to our interiors, our subjective life, as contrasted to the objective domain of material events and objects. Our spirituality is reflected in the values and ideals that we hold most dear, our sense of who we are and where we come from, our beliefs about why we are here—the meaning and purpose we see in our lives—and our connectedness to each other and to the world around us. Spirituality also captures those aspects of our experience that are not easy to define or talk about, such as inspiration, creativity, the mysterious, the sacred, and the mystical. Within this very broad perspective, we believe spirituality is a universal impulse and reality.[11]

When definitions become this broad, distinctions between religious and nonreligious activities become blurred, and the data can then be easily misinterpreted or skewed. Since the terms "spiritual" and "religious" are often used interchangeably, it becomes even more difficult to put a finger on the spiritual state of America. I would suggest that the idea of spirituality has become more popular in recent decades—people are certainly talking about it—but this does not mean that participation in religious and spiritual practices has increased. Instead, many researchers have found that religious and spiritual activity has been steadily declining over the past twenty years.[12]

Similar problems emerge in addressing experiences that might be called mystical or transcendent, but I would not necessarily consider any momentary uplifting event as falling into this category. Instead, I prefer William James's definition of mysticism in *The Varieties of Religious Experience* (published in 1902).[13] Such experiences are transient and rare, infused with ineffability and a noetic quality. "They are illuminations, revelations, full of significance and importance," said James, even though they may be difficult or impossible to articulate. James was aware that mystical experiences were often interpreted as emanating from a superior power or source, but he was also cognizant that such events could occur within secular contexts.

My research suggests that transcendent and mystical experiences can be traced to specific neural processes in the brain, and that they are available—and ultimately valuable—to anyone who seeks them, including secular individuals like Kevin. Religious experiences, therefore, should be considered a subset of transcendent experiences, since each religion tends to define transcendence in terms of its own system of beliefs.

With this distinction between mysticism and spirituality in mind, I asked Kevin to take Kass's test again. This time, the results were different. Kevin was presented with the following online commentary: "You have a healthy sense of well-being, but spirituality may not be a strong contributor. Perhaps you have not experienced your spiritual core or you are not giving yourself permission to recognize these experiences." Kevin found this offensive, for he felt that Kass was making the assumption that there must be something missing in a person's life if the person does not share Kass's view of spirituality. In other words, the investigator held certain religious biases that affected the commentary regarding various answers on the questionnaire. The difference between the two outcomes of Kevin's survey illustrates some of the problems that researchers face in trying to differentiate between spiritual, religious, and transcendent experiences. More significantly, it shows how religious and spiritual surveys are inclined to underestimate the number of nonreligious people in America.

According to professors Barry Kosmin, Egon Mayer, and Ariela

Keysar, authors of the largest research study ever conducted on American religious and secular practices, with more than 100,000 participants (and a follow-up study of 50,000 participants in 2001), there is a "wide and possibly growing swath of secularism among Americans," which is "frequently ignored by scholars and politicians alike."[14] They estimated that in 1991, 13 million Americans were nonreligious or secular; and that by 2001, the number had grown to nearly 28 million. That's a 110 percent increase in ten years. However, they calculated that fewer than 1 million Americans (0.4 percent) claimed to be atheists—i.e., disbelievers in God.

Looking at all the studies and surveys, it is difficult to accurately identify the religious and spiritual temperament of Americans. Depending on how the surveys are taken, what questions are asked, how they are asked, and how the terms are defined, you get contradictory results. And, as we shall see in Chapter 10, statistics can be easily manipulated to suggest the opposite of what is actually true.

Can Fantasies Heal?

Kevin does not believe in God, so I was bewildered when he told me that one of his favorite healing meditations involved an image of God. "I like to visualize the image that Michelangelo painted on the Sistine Chapel ceiling, of a compassionate wise old man with a flowing white beard. I imagine that I am being filled with a healing white light that enters my body from above." Kevin informed me that he had originally learned about this technique from Carl Simonton's research with cancer patients. Similar guided imagery is used in various psychotherapies and spiritual healing groups, and there is substantial evidence that guided imagery has psychological and physiological benefits, especially when used to treat pain, anxiety, and depression.[15]

For Kevin, God was merely a fantasy, but this practice presented a unique opportunity to see how Kevin's brain might process such an image. We know that if you imagine yourself eating a fudge brownie, you can taste it, because there are parts of your brain that do not distinguish between imagination and reality. But since Kevin doesn't believe in the reality of God, different circuits should be activated. As I have pointed out in previous chapters, a repeated focus on a specific image or concept tends to make it seem more real, but this does not seem to be the case with Kevin. "No matter how long I meditate, I never get the sense that God is real." The question naturally arises: could some people be born with a biological inclination toward spirituality, and others not? Recent genetic research points to this possibility.

Are Atheists Lacking a Spiritual Gene?

Accumulating evidence suggests that genetic factors may account for a substantial percentage of the individual differences in religious attitudes, interests and values.[16] In his book *The God Gene,* Dean Hamer, director of the Gene Structure and Regulation Unit at the National Cancer Institute, argues that spirituality is an instinct and that spiritually inclined people—specifically, those who claim to have self-transcendent experiences—are more likely to have the gene, VMAT2, that codes for a specific receptor in the brain.[17] Others have suggested that a spiritual or self-transcendent proclivity would probably involve multiple genes, including genes related to the dopamine[18] and serotonin neurotransmitter systems in the brain.[19] However, even if there is a genetic correlation with spiritual and transcendent proclivities, single genetic factors may have only a relatively small effect on a person's behavior, considering that our biology is governed by the simultaneous interaction of tens of thousands of genes. It is a huge speculative jump to say that a specific gene is responsible for a specific behavioral tendency or belief.

Although various studies support some relationship between genes and religious ideation, one's religious affiliation—along with the specific beliefs one chooses to embrace—is largely culturally and socially transmitted.[20] This means that genes do not turn a person into a Muslim, Hindu, or Catholic, for these are matters relating to child rearing, social norms, and an individual's freedom to choose. On the other hand, more general aspects of belief, such as religious fundamentalism, have been correlated with genetic factors.[21] Again, this does not mean that innate behaviors and attitudes cannot be changed, for many studies have shown that genetic tendencies can be overridden easily by cultural, environmental, and social factors such as education. For example, geneticists have found that a religious upbringing seems to inhibit—especially in boys and men—genetic tendencies that allow some individuals to express impulsive behavior and emotions that could potentially lead to destructive acts.[22]

One further possibility, which I have argued, is that the universal aspects of religion and spirituality—such as love, compassion, and feeling connected to something greater than the self—are a part of every human being. But as with any human trait, we each have varying predispositions and abilities. The result is that some people can feel highly spiritual while others do not.

The Atheist Brain at Work

After Kevin completed the interviews and questionnaires, we took him into a hospital examination room, where he rested quietly for ten minutes. We then injected him with a radioactive tracer. Ten minutes later, we took him for his baseline scan. This takes about forty-five minutes, during which time Kevin fell asleep. Falling asleep is not unusual, and Kevin had had a difficult time sleeping the night before. "Too much excitement," he explained.

When the baseline scan was complete, I compared it with the scans we had taken of the nuns and Buddhist practitioners. I knew that any conclusions I might draw would have to be considered cautiously, since Kevin was the first and perhaps the only atheist subject we'd be analyzing. After all, how many atheists do you know who meditate to an image of God? However, the argument has also been made that the data from individual case studies are as important as the information gathered from group studies, for they can highlight qualities that are unique within an individual's brain. Large studies tend to generalize data by excluding anomalies and extremes that may have relevance in studying the nature of the human mind. For example, if you were conducting a full-scale study on intelligence and creativity, you'd normally exclude statistical extremes. In essence, you'd be eliminating the Einsteins and Mozarts, since qualities of genius would be unusual compared with the norm. By studying exceptional individuals at both ends of the spectrum, we can begin to map a fuller range of human potential.

In several significant ways, Kevin's baseline scan turned out to be different from our other participants' scans, for he had higher activity in the prefrontal cortex than either the Buddhist meditators or the nuns. Frontal lobe activity plays an important role in mediating attention and controlling emotional feelings, and Kevin's brain seemed to be functioning in a highly analytical way, even when he was in a resting state.

If we make the assumption that atheism is a learned attitude that goes against the general beliefs of society (which would certainly be the case for Kevin), then I would argue that it takes a lot of cognitive work to embrace an atheistic point of view. Furthermore, I suspect that a significant increase in frontal-lobe activity would slow down neural activity in those parts of the brain that have a biological propensity to perceive alternative dimensions of reality. Kevin's personal experience confirms this possibility, for he told me that although his mind was relatively quiet as we did the baseline scan, he was still filled with thoughts about the experiment. "My mind is going all the time, thinking and imagining all kinds of things. My main reason I meditate is to turn the damn thing off."

Kevin also knew how controversial this experiment would be; and like most people, he was somewhat concerned about having his beliefs made available to others. Even though he knew we would take all possible steps to ensure his anonymity, going against the social norm, as I explained in Chapter 6, stimulates a neurological impulse to hide opposing views.

The nuns also had expressed excitement about participating in a study of prayer, yet they did not show the degree of activity seen in Kevin's scan. First of all, prayer is a widely accepted behavior, especially for a nun, so there would be no social dissonance to confront. Furthermore, the activity they were going to engage in complemented their system of beliefs, which is maintained by the frontal lobes. For an atheist, focusing on an image that contradicts one's beliefs could evoke two types of neural responses: first, an increase of negative emotional activity in the limbic system, which would potentially slow down frontal-lobe activity; second, an increase in frontal-lobe activity that would maintain a framework of disbelief. This is what we might be seeing in Kevin's highly active frontal lobes, for a predominance of frontal-lobe activity can also suppress the wider range of positive emotional experiences that I believe are essential to embrace spiritual perceptions such as God.

Compared with the nuns and Buddhists, Kevin also had lower activity in the hippocampus and right caudate, which are both associated with emotional responses. This might suggest that people who do not believe in God may have a decreased emotional range, at least when encountering religious stimuli. Kevin said, however, that he is often filled with a sense of peace and awe when entering churches and other sacred places. At first, his statement seemed contradictory, but then I discovered that whenever he travels, Kevin spends a great deal of time visiting art galleries, museums, and historical buildings. This suggests that Kevin's emotional reaction is based on nonreligious cues, such as the architectural beauty of a building or the aesthetic quality of a work of religious art. It isn't the religion that turns him on; it's the aesthetics.

Unusual Thalamic Asymmetry

In his baseline scan, Kevin also had substantial asymmetry in the thalamus (the left side being more active than the right). This was very similar to the asymmetry we found in the nuns, Buddhist practitioners, and Pentecostal practitioners. We typically do not see it in the "normal" population. Since the thalamus is a key relay of neuronal information in the brain, I have hypothesized that the asymmetry might be associated with long-term meditation processes. Since Kevin has been meditating for nearly thirty years, his resting scan supports this hypothesis. However, it is also possible that Kevin was born this way. If so, the inborn trait might explain why Kevin is "driven" to explore religious themes and spiritual practices, even though his other cognitive processes reject a spiritual cause. Long-term longitudinal studies would be needed to assess the validity of either perspective. Thus we are decades away from making more definitive statements about the biology of spiritual experiences and beliefs.

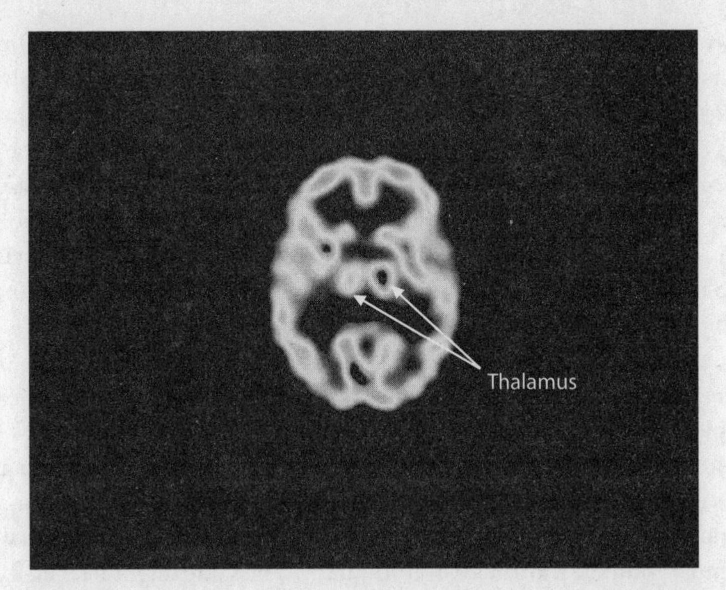

My findings also suggest that thalamic asymmetry may be associated with a predisposition to have powerful experiences while meditating, and as our experiment progressed, I discovered that Kevin could indeed evoke altered states of consciousness that had a powerful effect on his brain. For the religious person, this helps to validate the reality of a spiritual realm, and to keep that sense alive throughout the day. Thus, for the believer, this ongoing thalamic activity provides a more realistic sense of his or her faith.

For an atheist, a powerful transcendent experience might reinforce the belief that altered states of consciousness can be generated within the mind. For some people, such experiences may have only fleeting or entertainment value; but for others, radical shifts of consciousness have led to different fields of study. For example, Abraham Maslow's peak experiences became the focus of his academic work, which led to the founding of the American school of humanistic psychology and the establishment of the *Journal of Transpersonal Psychology* to investigate the therapeutic relevance of psychospiritual experiences.

Many atheists find themselves drawn to Buddhist practices for two primary reasons: first, they like the experiences that a meditative state evokes; and second, many forms of Buddhism are purely nontheistic, making no reference to deities or a supernatural realm. Instead, the focus is on quieting all thoughts and feelings, or on delving more deeply into one's psychological self. Thalamic activity and asymmetry may simply enrich the experience for such people by deepening the psychological reality of their beliefs. In Kevin's case, his meditation and peak experiences may have created his insatiable curiosity about the nature of the human mind. As he has often said to me, "I can't stop asking questions about the nature of people and the world."

In the clinical literature, people who practice intense forms of meditation and prayer sometimes report significant changes in their emotional life. Many people, for example, feel that they become more sensitive to the suffering of others because they see all life as interconnected. If prayer and meditation do increase our emotional and perceptual sensitivity, then it makes sense that parts of the thalamus would be more active, continually. The same might be true for highly creative and imaginative people who often see, feel, or hear dimensions of reality that others do not perceive. Kevin's asymmetric thalamus might be helping to guide him toward moral and ethical ideals that are similar to those encouraged by religious institutions.

Meditating on the Image of God

After we completed Kevin's resting (or baseline) scan, we returned with him to the examination room, where he would perform his "God" meditation. His technique bore many similarities to that of the nuns and Buddhists, allowing us to use the same protocol we had developed for them. The only significant difference was that Kevin tends to meditate only for as long as it takes him to enter a pleasurable altered state—about ten to twenty minutes—whereas the nuns and Buddhist practitioners spent nearly an hour in their meditative states. So the biggest issue for me was making sure that the tracer injection would be given at the appropriate time. This is important, since my research has suggested that there should be continual changes in the brain over the course of a meditation session.

We decided that Kevin would raise his finger when he felt that he was first entering an altered state. He signaled to me in less than ten minutes. I waited several minutes more before injecting him with the tracer, to ensure that the act of raising his finger would not distract him from his meditative state. Several minutes later, the experiment was complete.

Kevin was shocked when I told him the time, for he thought that nearly half an hour had elapsed. I asked him how he felt, and he exclaimed, "Marvelous! I haven't had such a powerful experience in years." He stated that his whole body surged with energy, and I noticed that his hands shook slightly as I led him back to the scanning room. Later, I asked him for a detailed description of his experience:

> I was fine when we first began, and had no trouble envisioning God's energy filling me with light and energy. I started to feel very, very relaxed, which is when I asked for the injection. But then I found myself having difficulty keeping my focus on God, or on the healing energy itself, and a series of thoughts intruded. I thought I was going to mess up the experiment. This evoked anxiety, so I increased my breathing by deeply inhaling and exhaling—usually this calms me down. Slowly I began to experience waves of energy and my whole body came alive, like light was shooting through me—very different from the warm healing light energy I usually visualize. Normally, I feel like I'm in control of the visualization, but the experience I was having seemed like it was directing itself. It almost felt overwhelming, and not unlike the beginning of those couple of psychedelic experiences I had in college. Then I lost track of everything. I didn't even have the sense that I was in a hospital room, or even in my body. And then my old self interrupted the experience, and I suddenly wished that I had postponed the injection because I was just beginning to enter an even more intense altered state. Then I kept bouncing back and forth between my thoughts and a sense of immersion, like riding an ocean wave. It was intense and pleasurable, and I didn't really want to stop when you touched my arm to signal that the experiment was complete, because I felt that I was just getting into the experience. Still, I felt pretty happy, and thrilled that something different had happened during the meditation.

There were a number of interesting findings on Kevin's meditation scan. One striking feature—quite different from what we found with the nuns and Buddhist practitioners—was that his frontal lobes were not particularly activated; in fact, they were mildly decreased during the meditation although in an asymmetric way. Usually the need to focus attention away from distracting stimuli results in increased activity in the frontal lobes, especially the prefrontal cortex (PFC). But Kevin, instead of focusing on God, had switched his concentration to his own vigorous breathing. This, various studies suggest, can account for a decreased activity in the PFC. According to one study at Harvard, slower breath meditations, which are aimed at relaxation, should increase PFC activity.[21] This is what we found in our studies of Buddhist meditation and the centering prayer. Vigorous breathing, however, when continued for many minutes, can even trigger hallucinogenic experiences, and that capability may account for Kevin's comparing his meditation to past psychedelic experiences. The lack of frontal-lobe activity suggests that when Kevin shifted his focus from God to breathing, the cognitive processes associated with any system of conscious belief would be suspended. This would be similar to what happened with the Pentecostal practitioners, who also showed decreased activity in the frontal lobes when speaking in tongues and dancing (which would also increase their breathing rate). In such an altered state, unusual auditory, visual, and tactile phenomena can be experienced.

Another part of the frontal lobe, the orbital frontal cortex, was slightly activated in Kevin. This area is known to be involved in tracking visual stimuli that have a positive emotional content. This may relate to Kevin's image of light shooting through his body, which triggered his feelings of elation. This same area of the brain was also activated in the nuns and Buddhists during their practices, and they too felt positive emotions about the various internally generated images and thoughts relating to their meditation.

Wanting to Believe

Although SPECT scans cannot show moment-to-moment changes in neural activity, it is reasonable to assume from Kevin's description—and from what we know about the neural processes involved when a person is facing ambiguous situations—that Kevin's brain was experiencing a form of cognitive dissonance. Kevin told me that he really wanted to immerse himself fully in the image of God during our experiment, but failed. In that moment, Kevin created a cognitive problem for himself; and as we saw in the brain scans of people facing moral problems (Chapter 6), such situations set off a flurry of conflicting feelings and thoughts. The fact that there was asymmetrical activity between the left and right lobes of the frontal cortex may reflect a conflict between his intention to "believe" and his ingrained sense that God was a fantasy.

Similar problems arise whenever we confront beliefs that are in opposition to those we personally cherish. Certain parts of the brain respond in a dualistic either-or way: either we are wrong and need to change our beliefs (this is biologically difficult for us to do), or the other person and his or her beliefs are wrong. In either case, we will tend to feel strong emotions to support our own beliefs, and thus we will try to denigrate the other person's beliefs. Ultimately, this can lead to animosity between individuals with differing beliefs, especially when the beliefs are about crucial life issues involving politics, nationality, and religion. In Kevin's case, the conflict was in his head as two cognitive ideas collided: "I want to believe" and "I don't believe." One solution to frontal-lobe dissonance might be to decrease the activity in both lobes. Intense breathing would achieve this.

With a decrease in frontal-lobe activity, one would expect to see increased activity in the emotional centers of the brain, corresponding to Kevin's experience and the experiences described by our Pentecostal subjects. In such cases, the emotional impact can stay with the person for days or even weeks. Kevin later reported that he remained in his energized state for days.

Ecstasy without Religion

Kevin's scan—just like the scans made during prayer and meditation—showed a decrease in his parietal lobes: the orientation area of the brain. He also felt a loss of his sense of self and a loss of his perception of space and time, a common occurrence in mystical, transcendent, and spiritual experiences. This supports the notion that there is a wide range of experiences in which the self and its relationship to the world can be altered. However, it is the interpretation of these experiences that ultimately leads to our beliefs about God, reality, and ourselves. For Sister Sarah, the timeless place is divine; to Kevin, it is simply a function within the brain.

A religious conservative might argue that Kevin simply did not have the strength or conviction to believe in the reality of God, and thus failed to achieve the kind of experience the Franciscan nuns did. An atheist, however, might point out that it is the strength of one's belief that creates the illusion of God. What we do know is that the stronger the experience, the more real it appears to be. Furthermore, we are all inclined to impose our belief systems on the perceptual mechanisms of the brain. We feel what we believe, and we believe what we feel, and this neurological loop gives us our sense of what is real. And the stronger the emotional impact, the more deeply these circuits will be embedded in the neural memory of the brain.

Altering Perceptions of Reality

For Kevin, the concept of God was just that: a concept, not reality. Because of this, the brain activation patterns during Kevin's meditation do not seem to be as strong as those in more spiritual individuals. The lower brain activation may also be related to the shorter time he spent meditating, compared with the other practitioners. It seems likely that the longer a person meditates during a given session, the greater the changes in the brain will be, so it may take close to sixty minutes for one to get the full perceptual and emotional impact of certain objects of contemplation. Had Kevin meditated for that long, I suspect that we would have seen more dramatic changes in his brain. I also suspect that intensive meditation retreats that last for days, and sometimes weeks or months, would also have profound effects on the overall functioning of the brain. The longer the process, the more intense the experience, and the more likely it is to change your perception of reality.

Distracting feelings and thoughts can interfere with contemplative practices. While highly proficient meditators can theoretically screen these out, less proficient meditators may be substantially influenced by them. For instance, Kevin admitted that he became anxious during his meditation. It may be that this anxiety altered his experience, and thus his brain, in unintended ways. Many meditation practices aim to diminish strong emotions which are believed to be destructive if they interfere with the ability to analyze real-life situations and live "in the moment."

The neural circuits that connect the higher parts of the brain to the thalamus, the limbic system, the parietal lobes, and ultimately the brain stem and body, form a complex network that creates, activates, and suppresses different beliefs at different moments. Our studies are beginning to show that each system of belief, and each form of meditation, activates a unique pattern of neural activity that changes the way we perceive reality.

Meeting the Atheist at the End of the Tunnel: A Near-Death Experience

Karen (a pseudonym) is a fifty-year-old social worker who has devoted most of her life to fighting all forms of discrimination. She is also the founder of a national free-thought society that takes a strong stand against prayer in public school, government-sponsored invocations, religious testing, and discrimination against atheists. Once, while recovering from a surgical procedure in the hospital, she suddenly felt herself floating over the nurses' station, where she observed someone looking at a fashion magazine. She tried to tell the nurse that there was a woman in a bed who needed help, but when she looked down, she saw that the woman was herself, slowly dying of suffocation. The next thing she recalls was walking through a tunnel of light, and there, at the end, was her atheist uncle, the person she credits with teaching her how to think critically. "Go back," he commanded. "You still have work to do."

Karen's next memory was of being revived by the nurses. She had nearly died. On her release from the hospital, she read some near-death literature, and she concluded that the incident was a purely naturalistic event caused by oxygen deprivation (one of many hypotheses that have been given to explain the tunnel effect).

When asked if the experience had in any way transformed her thinking or her life, Karen said that it had eliminated her fear of death, for now she believed that the process of dying would most likely be positive and painless. She didn't have any proof, but the experience made that belief feel real. William James would define Karen's experience as truly transcendent, but it did not change her fundamental disbelief in a spiritual realm.

Near-death experiences have been reported in all cultures. Some scientists believe they occur because of the loss of blood flow and oxygen to the brain, from drug effects, or from other neurochemical processes. But others believe that these explanations are incomplete, and that there is little evidence to explain the experiences. Regardless of their actual cause, near-death experiences are powerful neuropsychological events that often deepen fundamental beliefs about life; and such an experience itself usually becomes imbued with lasting personal meaning. However, near-death experiences are not always positive. At least one-third of those who have reported them feel traumatized by them.

Enlightenment without God

Individuals who engage in ritualistic practices such as meditation are often seeking some form of insight or personal transformation. Yet more secular endeavors, such as intense study or engagement in art or sports, may elicit similar changes. Mihaly Csikszentmihalyi, who has examined flow states and the psychology of optimal experiences, sees little difference between spiritual forms of meditation and meaningful or pleasurable activities that fully absorb a person's attention. When this happens, one's sense of self is temporarily lost. This, Csikszentmihalyi argues, evokes an experience of self-transcendence that enriches one's quality of life:

> When a person invests all her psychic energy into an interaction—whether it is with another person, a boat, a mountain, or a piece of music—she in effect becomes part of a system of action greater than what the individual self had been before. . . . The Self that is part of it expands its boundaries and becomes more complex than what it had been. This growth of the self occurs only if the interaction is an enjoyable one, that is, if it offers nontrivial opportunities for action and requires a constant perfection of skills.[24]

When an artist is inspired to paint or make a sculpture and succeeds in capturing an ineffable aspect of the subject, the artist's rapturous satisfaction may indeed be very similar to the transcendent experiences described by mystics and saints. But brain-scan technology is very new, and it will take many years of experimentation before we can compare the neurological processes that are involved when we elicit peak experiences.

Transcendent states, in William James's definition, are not reproducible, and each one results in an increased insight into life. In this sense, most meditative practices are designed to reinforce previous experiences and beliefs, rather than search out new ones, so the meditative state would not be considered transcendent. Practice makes one a better meditator, in much the same way that practice makes a person a better musician or surgeon. But transcendence for a great artist or philosopher may be a point at which a new illumination supplants old beliefs.

The Pentecostal practice of speaking in tongues, however, seems to come closer to James's definition of transcendence, since practitioners often use the experience to change some aspect of their lives. Kevin's experience in our lab also may have captured certain elements of transcendence, for he reported to me, a year after our experiment, that he no longer meditates on God. "I found the experience a little too intense," he said. "And since I don't really believe in God, I thought it might be more prudent to focus on those goals I truly want to embrace: happiness, peace, and compassion." Kevin chose to take a different path that was more consistent with his secular beliefs about the universe.

Atheism and Health

Many studies have attempted to correlate religion with health, but none has clearly shown that atheism is an unhealthy belief system. By itself, a belief system cannot predict whether an individual will be happy or healthy. For example, a religious person who struggles with the tenets of his or her religion is likely to experience anxiety and stress, whereas a nonreligious person who derives great pleasure from secular beliefs will probably experience a high degree of satisfaction with life. Happiness is generated from multiple factors that involve social and family life, physical and emotional health, satisfaction with work, intellectual stimulation, and even altruistic pursuits. Religious beliefs are but one part of this complex interaction.

In a study conducted at the University of Illinois, 222 undergraduates were screened for happiness using several assessment filters. The researchers reported the following:

> We compared the upper 10% of consistently very happy people with average and very unhappy people. The very happy people were highly social, and had stronger romantic and other social relationships than less happy groups. They were more extraverted, more agreeable, and less neurotic, and scored lower on several psychopathology scales of the Minnesota Multiphasic Personality Inventory. Compared with the less happy groups, the happiest respondents did not exercise significantly more, participate in religious activities significantly more, or experience more objectively defined good events.[25]

Thus, happiness is not necessarily related to one's religious or spiritual beliefs. The most important element, according to these researchers, was maintaining a network of good social relationships. What, then, is one to make of the hundreds of studies supporting the notion that religious involvement enhances one's emotional and physical health? There are four important issues to consider in reviewing studies on religion and spirituality.

First, we have to recognize that the beneficial impact of religion, though statistically significant in many studies, is often small. More important, many factors not relating to spirituality per se are also involved. For example, religions offer social interaction; meaning in life; rules against unhealthy behaviors such as excessive drinking, smoking, or promiscuity; and a variety of psychological coping mechanisms. However, a person can also have access to these healthful elements through nonreligious groups. When religion does provide these healthy elements, it can be very beneficial; but the question is whether there is something intrinsic to religiousness itself that makes it healthier than other belief systems. In this sense, no one has yet designed the "perfect" study to account for all the variables that are involved in religious activities and personal health.

Second, most studies involve self-reports, which tend to be optimistically biased. In other words, nearly everyone tends to believe that the activities he or she chooses to engage in are beneficial, even when they are shown to be otherwise. People will also tend to ignore behaviors known to be unhealthy. Alcoholics, for example, will underreport their drinking,[26] and nearly half of all adolescents will deny that they have ever had a sexually transmitted disease, even though their medical records state otherwise.[27]

Third, the wording of many studies biases the outcome. This is not done deliberately; it's just one of the problems researchers face when trying to define their terms. The best a research study can do is point to a possible answer, rather than an absolute truth.

Fourth, researchers often include experiences such as optimism, pleasure, peacefulness, forgiveness, and kindness as indicators of spiritual, mystical, and transcendent states. In fact, these attitudes promote health in both religious and nonreligious individuals.

When you take all these influences into account, it is difficult to argue convincingly that, on an individual basis, nonreligious people are less happy and less healthy than those who believe in God. In fact, according to the research of Phil Zuckerman, a professor of sociology at Pitzer College, "In sum, countries marked by high rates of organic atheism are among the most societally healthy on earth, while societies characterized by non-existent rates of organic atheism are among the most destitute. Nations marked by high degrees of organic atheism tend to have among the lowest homicide rates, infant mortality rates, poverty rates, and illiteracy rates, and among the highest levels of wealth, life expectancy, educational attainment, and gender equality in the world."[28] Citing the findings from contemporary research, Zuckerman concludes that "societal health causes widespread atheism, and societal insecurity causes widespread belief in God."

Perhaps the most important thing to keep in mind is that all statistical surveys have built-in limitations. As the largest independent social research institute in Britain points out, pollsters incorrectly assume that respondents understand their questions in the ways the pollsters intended, and they also assume that all respondents answer in similar ways.[29] With this in mind, let's take a look at some of the more respectable surveys, and the problems they raise.

In a study conducted by the Barna Group, atheists had a lower divorce rate than religious groups,[30] but according to the ARIS study, which included over 100,000 participants, atheists also had the lowest marital rate.[31] If you believe in the sanctity of marriage, poor statistical logic might lead you to the conclusion that you should give up your religious beliefs the day after your wedding in order to stay married.

In another reputable study, atheists reported higher levels of stress and less satisfaction in life than evangelicals.[32] Does this mean that strong religious beliefs are healthy for you? On one hand, this might be the case, but on the other hand, a recent study by the Mayo Clinic reported that compared with atheists and agnostics, highly religious people had more obsessional symptoms, showed more intolerance for uncertainty, needed to control thoughts, and had an inflated sense of responsibility.[33]

In a study that statistically analyzed the beliefs of people in sixty-six countries,[34] religious people tended to trust others more, including the government, and were less willing to break the law. However, they also tended to be more racist, showed less concern for the rights of working women, and expressed greater intolerance toward other religious groups. Buddhists, by the way, showed the greatest tolerance toward others. Atheists were more tolerant of others who held different beliefs, but were less trusting of the government and more willing to break the law.

The problem is this: opposing groups will selectively use data to support their point of view. Proponents of religion might say that atheists are thieves; and atheists might counter by pointing to the racism that religious groups generate. Both sides, however, would have adapted the findings to their own beliefs. When you look at the data from a broader perspective, it seems clear that each individual is free to decide what is right or wrong, and that everyone, no matter what his or her religious orientation may be, has certain weaknesses and strengths. Human beings are full of faults. If religion helps some improve, great; and if being an atheist means fighting religious injustices, then this outlook also makes a contribution to our society. It all depends on your innermost beliefs and how you choose to manifest them in the world.

Does Education Weaken Religious Belief?

In religious and secular groups, there is a common belief that education tends to weaken religious ideas, but what is the actual evidence? According to one Harris Poll, education may have only a minimal effect on spiritual and religious beliefs. Ninety-two percent of Americans with no more than a high school education believe in God. With each year of college education, the percentage drops, but not by much. Ninety percent of college graduates continue to believe in God, and 85 percent of postgraduates believe. That's less than a 10 percent drop with higher education. Beliefs in miracles, however, falls off by nearly 20 percent (from 89 percent to 72 percent). Belief in hell and the devil decreases by nearly 30 percent. Still, more than half of the postgraduates polled maintained these religious beliefs.[35]

On the other hand, among scientists, religiosity has been steadily declining for the past 100 years. In 1914, James Leuba surveyed 400 noted scientists and found that 53 percent did not believe in a personal God.[36] When he repeated his study twenty years later, Leuba discovered that 68 percent no longer believed in God. In fact, only 15 percent continued to believe in a personal God.[37] In 1997, in an effort to improve on Leuba's strategy, Edward Larson and Larry Witham sent questionnaires to 517 members of the National Academy of Sciences, the most prestigious science organization in the world. Of the half that responded—the group mainly included biologists, mathematicians, physicists, and astronomers—72 percent did not believe in God, and only 7 percent believed. Another 21 percent expressed religious doubt or took an agnostic position.[38] By an overwhelming percentage, scientists show the least interest in religion.

However, a very different picture emerged from a preliminary survey put together by Elaine Ecklund at Rice University. She questioned a much broader group of faculty members from various universities that included researchers in the natural, social, and political sciences. Only 34 percent of the respondents said that they did not believe in God, less than half the percentage in the survey by Larsen and Witham. However, Ecklund focused on "spirituality," and although this term was undefined, the study suggested that "as people move away from traditional religious beliefs, they continue to maintain an interest in personal spiritual matters."[39]

When we put all these studies together, they still suggest that at the highest levels, science and education undermine traditional religious beliefs, including specific beliefs about God. However, the reasons for this remain obscure. Perhaps individuals working at these levels find the biological and astronomical evidence more compelling than the arguments put forth by religion, or perhaps the findings simply reflect yet another shift in the ever-changing spiritual landscape of America. Some observers, particularly religious traditionalists, might argue that it is arrogance to presume to "know" the world through science and human investigation.

However, it is important to remember that at least one-fourth of our brightest scientists continue to allow for the possibility of God, with all the paradoxes it brings. As Stephen Hawking wrote in *A Brief History of Time,* if someday we should ever find a complete and unified theory of the universe:

> Then we shall all, philosophers, scientists and just ordinary people, be able to take part in the discussion of the question of why it is that we and the universe exist. If we find the answer to that, it would be the ultimate triumph of human reason—for then we would know the mind of God.[40]

According to the distinguished professor Henry F. Schaefer III, Hawking has made it very clear on numerous occasions that he is not an atheist,[41] but when a scientist does not take a nonreligious stance, public controversy bursts forth, from both sides of the aisle, and sometimes the facts are lost. Darwin, for example, was not an atheist, as many people believe. In his brief autobiography, he wrote that he found it extremely difficult to think of the creation of the universe "as blind chance." He said that "the mystery of the beginning of all things is insoluble by us," and so he concluded, "I for one must be content to remain an Agnostic."[42]

In any case, atheism does not necessarily imply hostility toward religion or spiritual pursuits. Einstein, for example, rejected belief in a personal God while maintaining respect for religious principles. He believed that genuine scientific inspiration "springs from the sphere of religion."[43] Furthermore, he felt that religion had the capacity "to liberate mankind as far as possible from the bondage of egocentric cravings, desires, and fears,"[44] and that science was not equipped to do this.

The Neurochemistry of a Skeptical Brain

Neurological evidence suggests that the brains of believers and skeptics function differently. According to research neurologists at University Hospital in Zurich, Switzerland, when subjects viewed scrambled words and phrases on a screen, believers were much more likely than skeptics to see words and faces when there were none, but skeptics often didn't see words and faces that were there. However, when skeptics were given the drug L-dopa (used to treat Parkinson's disease) to increase dopamine levels in the brain, they were more likely to interpret scrambled patterns as real words and faces. The researchers concluded that believers use looser criteria for interpreting sensory information, and so are more likely to make unfounded inferences. These looser criteria may also explain why certain individuals are more inclined to form paranormal beliefs.[45] On the positive side, the researchers also suggested that higher levels of dopamine may be "a prerequisite of creative thinking."[46] However, I want to point out that this study also found that both skeptics and religious believers make significant mistakes in processing their perceptions of the world.

Interestingly, religious beliefs and spiritual experiences may be deeply influenced by a variety of neurochemical and hormonal interactions. For example, the study in Zurich suggests that dopamine may play an important role in spiritual experiences, and that religious practitioners may have higher levels of dopamine than nonreligious individuals. Other research has suggested that the balance of activity between the brain's left and right hemispheres could regulate a person's predisposition to spirituality or atheism.[47] The neurotransmitter serotonin, which regulates emotions, behavior, and thoughts, might also play a contributing role in spiritual experiences.[48]

Taken together, all these studies suggest that there is no one spot in the brain—no specific function or chemical balance—that makes us religious or atheistic. Our brain works as a whole, and thus all our beliefs are affected by every part of the brain. Furthermore, each system of belief and disbelief has strengths and weaknesses. As the study in Zurich implies, believers would have a tendency to see affirmations of spirituality (miracles, paranormal phenomena, etc.) where there are none, whereas nonbelievers would tend to dismiss significant findings in these areas.

On an even larger scale, such studies support my argument that all beliefs, by their very nature, tend to exclude contradictory evidence. We may even be born with individual biases hardwired into the structure of the brain.

Believing in Each Other

In conclusion, the evidence suggests that we should be very careful about making causal connections in the accumulating data concerning spiritual and religious beliefs. There are compassionate, creative atheists; and there are murderers who act in God's name. This suggests that it is important to judge people not only on their beliefs but also on the behaviors that arise from those beliefs. Spiritual beliefs—all beliefs, for that matter—reflect the way we choose to understand reality, and this is a unique experience for every individual. Truth, beauty, compassion, morality—all such ideals can be embraced by religious and nonreligious people alike.

Different beliefs can open the mind to possibilities previously undreamed of, and this open-mindedness can be best achieved by maintaining a compassionate dialogue between all sides of the spiritual debate, especially between scientific and religious views. I believe that this is what Einstein was suggesting when he said that "science without religion is lame, religion without science is blind."[49] Whether we are gazing through a telescope, or contemplating our soul, we all can marvel at the beauty and mysteriousness of the universe. It is in the nature of our brain to search for its deepest truths, and although we may never grasp truth in its entirety, it is our right, and our biological heritage, to try.

Chapter 10

Becoming a Better Believer

> Nothing is easier than self-deceit. For what each man wishes, he also believes to be true.
>
> —Demosthenes (384–322 B.C.)

IN 2001, IN THE CHRISTMAS ISSUE OF *BMJ* (FORMERLY THE *British Medical Journal*), Leonard Leibovici, a professor at the Rabin Medical Center in Israel, published a remarkable study on the healing power of prayer. Numerous scientific studies have reported that long-distance healing—praying for someone far away who doesn't know about it—can actually occur, and an equal number has shown that it does not occur. But Leibovici's experiment had an unusual twist: the prayers were being sent to patients who had been hospitalized five to ten years earlier, and had already been released from the hospital.

It seems impossible that one could influence the past in such a way. So you can imagine my surprise when I read in the papers that the outcome was positive. Of the 3,393 hospitalized patients who were diagnosed with an infectious blood disease, half became a control group (for whom no prayers were given) and the other half received prayers from a remote intervention group. In the control group, 30 percent of the patients died, but in the intervention group, only 28 percent died. More important, those who were prayed for had significantly shorter stays in the hospital, and a shorter duration of fever symptoms. On the basis of these findings, Leibovici, who has more than 100 scientific papers credited to his name, declared that "remote, retroactive intercessory prayer . . . should be considered for use in clinical practice."[1]

Within weeks of publication, the article had fomented controversy, and *BMJ* was flooded with responses from doctors, scientists, and professors around the world.[2] Andrew Thornett of Adelaide University Rural Clinical School in Australia pointed out that the effects on mortality were statistically insignificant, and that the other effects could be explained by natural rather than supernatural causes. Another commentary, from the physicians Shehan Hettiaratchy and Carolyn Hemsley, noted how a few abnormal results in the control could skew the statistical findings. But other medical reviewers felt that the experiment was well designed.[3] Basically, those who believed that thoughts could influence others tended to see the study in a positive light; those who didn't believe this thought that the study must be flawed. This suggests to me that certain unconscious biases were influencing how people evaluated the work.

In response to numerous queries challenging the seriousness of his intentions, Leibovici responded that he wrote the paper to demonstrate how certain ideas, when coupled with statistical methodology, could be carried out to absurd conclusions.[4] In a published response, Larry Dossey, well known for his medical research on the power of prayer, and Brian Olshansky, a medical professor at the University of Iowa wrote:

> Questions raised by intercessory prayer and distant healing are far-reaching, challenging basic assumptions about the nature of consciousness, space, time, and causality. Many consider these issues vexing and simply ignore them. . . . Rather than dismissing studies of prayer because they do not make sense or confirm our existing knowledge, we should consider them seriously exactly for this reason. In the history of science, findings that do not fit in often yield the most profound breakthroughs.[5]

The experiment, Dossey and Olshansky argued, went far beyond questions concerning prayer, for it raised the issue of whether consciousness itself can influence objects in the world. For decades, this has been an appealing proposition, and hundreds of studies have touched on such issues as telepathy, telekinesis, and distance viewing. Even though the statistical significance is slight, proponents have argued that the probability of such findings is hundreds, or even thousands, of times more than mere chance. For example, Dean Radin, the director of consciousness research at the University of Nevada, analyzed all the known experiments studying distant mental influence and concluded that the odds against such phenomena being chance were 1.4 million to 1.[6] And although the idea that one can influence past events goes against everything we know about the workings of the physical universe, an analysis of more than twenty controlled studies in which someone tried mentally to shape past events showed that the majority of those experiments also found statistical significance.[7] One counterargument is that few of these studies have been replicated.

One of the problems, as I see it, is not in the studies themselves, but in the way we draw our conclusions. This is particularly important when we use quantitative evaluations, because statistics, as I pointed out in Chapter 9, are not in themselves an accurate assessment of truth. They are given meaning by those who interpret them, and interpretations reflect unconscious biases and preferences generated by our brain. As Mark Twain famously said, there are "lies, damned lies, and statistics."

Damned Lies and Statistics

As I discussed in Chapter 4, one of the analytic functions of the brain is to quantify perceptions, a process that deeply influences our beliefs about the world. In this sense, statistics are tools for organizing information in ways that allow us to evaluate the relevance of collected evidence. But they are only a guide—showing values, comparisons, and relationships—and they can be easily manipulated to suit the purposes of the researcher.[8]

Let me show you how this works. Often, in medicine, you'll hear that a particular drug, procedure, or intervention will reduce your risk of a disease by 50 percent. On the surface, this would appear to be a powerful inducement for seeking the treatment. But the percentage is not given in any context. For example, male circumcision reduces the incidence of penile cancer by 50 percent, or to put it another way, uncircumcised males are twice as likely to develop penile cancer.[9] That's a 100 percent increase in risk. If you happen to be the parent of a newborn son, these statistics might encourage you to opt for circumcision. However, only two out of 100,000 uncircumcised boys will ever get this rare form of cancer. Circumcision does cut the risk in half, but the risk is only one ten-thousandth of 1 percent. Now we have a handful of percentages to choose from: a 50 percent decrease in risk, a 100 percent increase in risk, and one chance in 100,000 of ever getting penile cancer. Proponents of circumcision will cite one number; opponents will cite another. Add to this the fact that approximately one in 100 boys will have postsurgical problems, and you can see how different statistics concerning the same event can influence the choices we make. (It turns out that nearly half of the physicians performing circumcisions did not discuss the medical risks with the parents.[10]) Medical research depends on statistics to support various treatment modalities, but this does not mean that a procedure has significant value when the risks involved are taken into account.

There are so many ways to analyze statistics that one almost has to be an expert to know what certain statistics actually mean. And when someone turns them into colorful charts and three-dimensional graphs, the information appears more dramatic, and further distorts the data. To give you an example, I used my word processor to make two charts. Let's say that they represent rainfall over several months in six countries you are planning to go visit. In the smaller chart, the first three countries are labeled A, B, and C. In the longer one, the other three countries are labeled D, E, and F.

In the first chart, it appears that all three countries have dramatic changes in precipitation and a tremendous amount of rain; but in the second chart the precipitation seems less extreme. Country C is particularly hard to evaluate because part of it is blocked by the rain spikes in countries A and B. So, assuming that I wanted mild

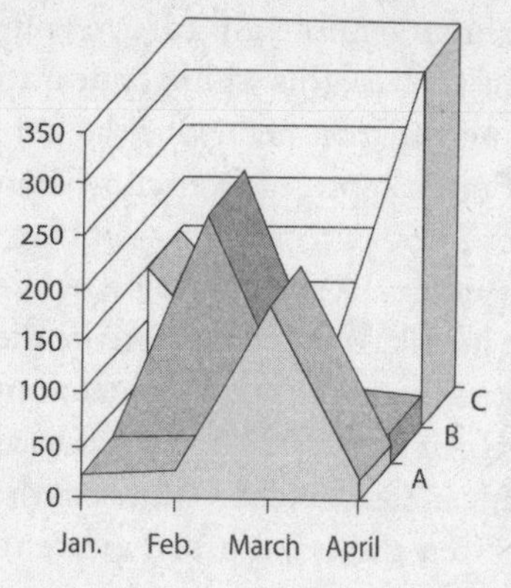

weather for my vacation, I'd probably choose country F; and if I couldn't get a decent hotel there, I'd opt for the flat season in D or E.

Here's the problem: the two charts contain exactly the same data. I actually made two copies of the same graph. I stretched one vertically, stretched the other horizontally, and changed the letters. (The computer, by the way, automatically changed the numbering on the left-hand side.) The information is the same, but the visual effect is vividly different. In the second chart you can even see the areas that were hidden in the first chart. Remember, country A is the same as country D, B is the same as E, and C is the same as F. My question to you is this: which chart gives an accurate picture of the rainfall? The answer is: neither one. They both give the same information, but the ambiguous way it is graphed affects our emotional reaction, and the first chart stimulates a stronger emotional response in the brain.

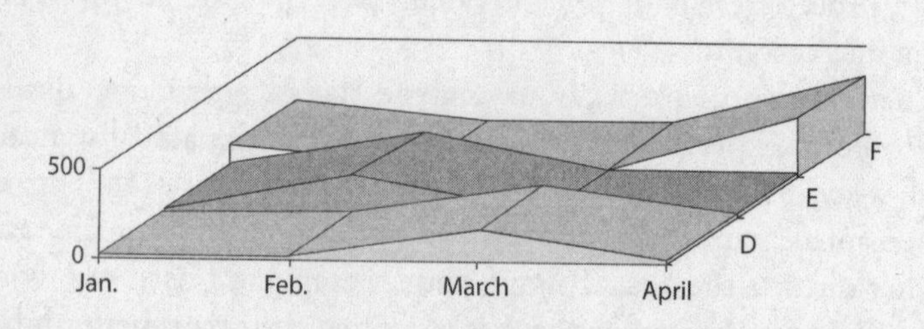

The creators of stock market charts are notorious for these kinds of manipulations; a one-month graph can show that your mutual fund has soared in value; a one-year graph might show that the same mutual fund has actually lost money. Dramatic charts attract more attention in a magazine, but at the expense of muddling the facts.

Remember that scientific research has inherent limitations, simply because it is impossible to control for all the variables involved.

So the best we can do is use our studies as indications and possibilities, rather than proofs. We propose hypotheses and interpretations, but science has difficulty identifying absolute truths.

What, Exactly, Is a Prayer?

In scientifically evaluating the effects of distant prayers, statistics are only one part of the problem. There are many other issues that need to be addressed. For example, if more people pray for a specific outcome, would that increase the effect of the prayer? Logically, it makes sense to say yes; but no one, to my knowledge, has tried to evaluate this factor. Imagine, if you will, a war between two religious groups. Will the side with the greatest numbers of people praying win? Such hypothetical situations may sound absurd, but in religious communities, questions concerning the power of prayer are debated every day.

In times of war, we pray for victory and we believe, or hope, that God is on our side. But if prayer can have an effect in war, or on the health of an ailing relative, then isn't it possible that prayer can work in other situations? Think of a football game, where two groups of fans are rooting for their team to win. Will more intense cheering help one side triumph over the other, and will cursing the opposing side have a negative effect on the players? Certainly, internal negativity can interfere with an athlete's performance, a student's grades, or a patient's health; but in the relevant studies, the people who are being cursed, cheered, or prayed for know that this is occurring. In double-blind studies of prayer, however, neither the patient nor the doctor knows that positive thoughts are being sent.

Furthermore, if praying for others does help them get better, what would happen if you prayed for someone to get ill? Here the ethical issues soar, for it might be immoral to send negative thoughts to uninformed subjects. However, if you don't believe that negative prayer could possibly have an effect, then there should be no problem in carrying out such an experiment. The catch is this: if the study ultimately does show that negative prayer works, then you have violated the rights of the patient.

Another major difficulty with such experiments is how to control for other thoughts that are generated in the brain of the patient or the person who is praying. I encountered this problem with Kevin in the study of atheism, since he had numerous distracting thoughts, any of which could have influenced the overall brain activity we were recording.

Studies of prayer raise other questions that have yet to be addressed. For example, not everyone prays in the same way. Is a parish priest's prayer less potent than the pope's? Is a Christian prayer as effective as a Hindu meditation? Does prayer work when you're praying for a person of a different religious faith? Would prayers by atheists have effects similar to prayers by believers? With regard to the patients, will prayers received by disbelievers have the same effect as prayers received by believers? One study actually touched on this issue, and the researchers found that a "marginally significant reduction in the amount of pain was observed in the intervention group," but only for those patients who believed that their problem could be resolved.[11] This suggests that if you do not believe in the possibility of healing, prayers will not have an ameliorating effect. But again, this tells us more about the internal state of the patient, and less about the physical effects of prayers received from others.

Finally, most studies of prayer fail to address what a prayer actually is. Is it just a thought, or is something more involved? Must the thought include reference to a particular deity or religion, or is the power of prayer based on the intensity with which one prays? How do you measure such qualities, and what methodological formats would you use? Ultimately, whatever results you get will still not tell you anything concerning the existence or nature of God.

What is current opinion concerning distant healing? An overview of all the research suggests "maybe." In 2000, in a systematic review of 100 clinical trials that included prayer, therapeutic touch, and other forms of distant healing, the researchers identified twenty-three studies that (in their opinion) had used adequate protocols. Thirteen of these had found small but statistically significant treatment effects; nine had found no effect; and one had found a negative effect.[12] In 2003, Edzard Ernst updated this study, including nine additional randomized trials carried out since 2000, one of which was a highly controversial and apparently fraudulent study at Columbia University[13] (infertile women who received Christian prayers were reportedly twice as likely to conceive over those who were not prayed for[14]). Even though these new studies found positive results, Ernst concluded that distant healing was no more effective than a placebo.[15] However, as I will argue shortly, the placebo effect, which is largely based on a person's belief system, can profoundly influence the healing process, and thus should not be used as evidence against the power of prayer.[16]

Twenty-seven Ways Our Brain Distorts Reality

The knowledge we glean from scientific studies depends largely on how we interpret the evidence. But interpretations are subject to the same rules that govern our perceptions of reality; they are filled with assumptions, generalizations, oversights, and mistakes. In the social sciences, these errors are referred to as cognitive biases; but as I have emphasized throughout this book, such biases are built into the perceptual and emotional as well as the cognitive mechanisms of the brain. By the time perceptual information reaches consciousness, each individual has transformed it into something new and unique. This reconstruction of reality is the foundation from which we construct all our beliefs about the world.

Logic, reason, and social consensus also play critical roles in shaping our beliefs; but as we have seen throughout this book, these factors also bias the way we understand the world. By recognizing these biases, we can become better thinkers, better researchers, and ultimately better believers. Over the last fifty years, researchers, scientists, psychologists, and sociologists have identified hundreds of cognitive, social, behavioral, and decision-making processes,[17] and I have gathered here twenty-seven biases I consider essential for evaluating our perceptions and beliefs about the world.

1. **Family Bias** We have a propensity to automatically believe information given to us by family members and close friends. Our brain has relied on these individuals throughout our life, and thus we tend to accept their word without checking the facts.

2. **Authoritarian Bias** We tend to believe people who hold positions of power and status. We give them more credence without checking their sources.

3. **Attractiveness Bias** We give greater credence to taller, more attractive individuals because the brain seeks what is aesthetically pleasing. People who make more eye contact are also more likely to be believed.

4. **Confirmation Bias** We have a tendency to emphasize information that supports our beliefs, while unconsciously ignoring or rejecting information that contradicts them. Since beliefs become embedded in our neural circuitry, contradictory evidence often cannot break through the existing connections in the brain.

5. **Self-Serving Bias** In conjunction with the confirmation bias, we also tend to maintain beliefs that benefit our own interests and goals.

6. **In-Group Bias** We unconsciously give preferential treatment to other members of our group and rarely question their beliefs, because our brains are wired to seek conformity with others.

7. **Out-Group Bias** We generally reject or disparage the beliefs of people who are outside our group, especially when their beliefs differ markedly from our own. In addition, we have a biological propensity to feel anxious when encountering people from different ethnic and cultural backgrounds, even if they are members of our group.

8. **Group Consensus Bias** The more other people agree with us, the more likely we will be to assume that our beliefs are true. Conversely, the more people disagree with us, the more likely we'll be to suppress and doubt our own beliefs, even if they are correct.

9. **Bandwagon Bias** This reflects our tendency to go along with the belief systems of whatever group we are involved with.

The more people we are surrounded by, the more likely we'll be to modify our beliefs to fit theirs.

10. **Projection Bias** We often assume, without checking, that other people in our group have similar beliefs, have similar morals, and see the world in similar ways. The Central Intelligence Agency refers to this bias as the "everybody thinks like us mind-set" and considers it one of the most dangerous biases a person can have—because different cultures, and different personality types (such as terrorists) don't think like us.

11. **Expectancy Bias** When looking for information, or conducting research, we have a propensity to "discover" what we are looking for. In medicine, double-blind studies try to eliminate this pervasive bias.

12. **"Magic Number" Bias** Numbers influence our beliefs because of the strong quantitative functions of the brain. The larger and more dramatic a number is, the greater emotional impact it will have, and this, in turn, strengthens our trust in the information being quantified.

13. **Probability Bias** We like to believe that we are luckier than others, and that we can beat the odds. (Depressed individuals, by contrast, tend to believe the opposite.) This optimism is also known as the gambler's bias. If you flip a coin that comes up heads nine times in a row, most people will bet a lot of money that the next flip will be tails. Of course, the probability remains the same for every flip; you always have a fifty-fifty chance of being tails. We also maintain magical biases that are carried over from childhood. Thus many adults, especially gamblers, keep special items (a four-leaf clover, a rabbit's foot, a coin) to help bring them luck.

14. **Cause-and-Effect Bias** Our brain is predisposed toward making a causal connection between two events, even when no such connection exists. If you take an herbal remedy and your cold disappears, you'll attribute the cure to the remedy, even though dozens of other unrelated factors may be involved.

15. **Pleasure Bias** We tend to assume that pleasing experiences reflect greater truths than unpleasant experiences, in part because the pleasure centers in the brain help control the strength of perceptions, memories, and thoughts.

16. **Personification Bias** We prefer to give inanimate objects lifelike qualities. We also tend to give ambiguous stimuli (shadows, indistinct sounds, etc.) human and animal-like forms. This perceptual and cognitive function gives rise to various superstitious beliefs.

17. **Perceptual Bias** Our brain automatically assumes that our perceptions and beliefs reflect objective truths about ourselves and the world. This leads to the old saying, "Seeing is believing."

18. **Perseverance Bias** Once we believe in something, we will continue to insist that the belief is true, even when contradictory evidence is offered. And the longer we maintain specific beliefs, the more ingrained they become in our neural circuitry.

19. **False-Memory Bias** Our brain tends to retain false memories longer than accurate memories. It is also easy to implant false memories in others if the circumstances are right and the information is plausible.

20. **Positive-Memory Bias** When reflecting on the past, we tend to recall events in a more positive and favorable light than they had when they first occurred.

21. **Logic Bias** We tend to believe arguments that strike us as more logical. We also tend to ignore information that doesn't make sense to us. As William James said, "As a rule we disbelieve all the facts and theories for which we have no use."

22. **Persuasion Bias** We are more likely to believe someone who is more dramatic and emotional when arguing a particular point of view. Our brain tends to resonate with great speakers, and we can get caught up in their emotions and their beliefs.

23. **Primacy Bias** We give more weight to, and remember more easily, names and information that appear at the top of a list.

24. **Uncertainty Bias** Our brain does not like uncertainty and ambiguity; thus we prefer to either believe or disbelieve, rather than remain uncertain.

25. **Emotional Bias** Strong emotions usually interfere with logic and reason. Anger tends to evoke the belief that we are justified and right; anxiety undermines such a belief; and depression obscures optimistic beliefs.

26. **Publication Bias** Editors of books, journals, and magazines prefer to publish work that shows positive outcomes, and to exclude work with negative findings. Thus a research project that shows no effect is less likely to be published than one finding positive effects. Another dimension of this bias is the propensity of readers to assume automatically that anything published is true, even when it appears in the tabloids.

27. **Blind-Spot Bias** Last, but not least, researchers have identified a blind-spot bias. Most people fail to recognize how many cognitive biases they actually have, or how often they fall prey to these biases.[18] Advertisers and politicians are very much aware of these blind spots, and they deliberately appeal to our biases to sell their products and ideas. To a certain extent, we all manipulate others to persuade them to embrace our own beliefs. Parents do so with their children, teachers with their students, researchers with their colleagues, and lovers with their beloved. Unfortunately, we often do this without consciously considering the other person's interests or needs.

The Foolish Brain

Our propensity for cognitive biases is ritually demonstrated once a year, when a large percentage of people go out of their way to test the gullibility of others. On April Fools' Day, millions of unsuspecting victims discover how prone they are to believing outrageous tales. In 1957, the British Broadcasting Corporation (BBC) showed a film of spaghetti being harvested from trees. So many viewers called up wanting to know how to grow their own that the station replied by telling them to "place a sprig of spaghetti in a tin of tomato sauce" and patiently wait for their harvest. In 1976, the BBC announced that the planet Pluto would have an unusual effect on gravity when it passed behind Jupiter, and that if you jumped at just the right moment, you would feel a strange but wonderful floating effect. Hundreds claimed to experience it. In 2002, a British supermarket advertised a genetically altered carrot that would whistle when properly cooked; this brought hundreds of customers to the store. And lest one think that only the Brits are gullible, the White House announced in 1996 that it had sold the Lincoln Memorial to the Ford Motor Company. Hundreds of outraged people called in to complain. On the same day, Taco Bell ran an ad in the newspapers claiming to have purchased the Liberty Bell.

The moral of these stories is obvious: don't believe everything you read or hear. And the neurological explanation for this is simple: our brain is calibrated to trust anyone who happens to be a "member" of our group or an authority figure. And so we are biologically biased to believe the magazines we buy, the news channels we select, and the people we personally like.

The CIA's War against Biases

If you want to become a better believer, the first step is to realize that every perception and thought includes a degree of bias, and thus every belief represents a compromise between the way the world really is and the way we would like it to be. This is such a difficult notion to accept that a special branch of the Central Intelligence Agency recently published a book—a sort of in-house training manual—to emphasize the fact that we are strongly biased toward perceiving an inaccurate view of reality. To quote:

> People construct their own version of "reality" on the basis of information provided by the senses, but this sensory input is mediated by complex mental processes that determine which

information is attended to, how it is organized, and the meaning attributed to it. What people perceive, how readily they perceive it, and how they process this information after receiving it are all strongly influenced by past experience, education, cultural values, role requirements, and organizational norms. . . . We think that if we are at all objective, we record what is actually there. Yet perception is demonstrably an active rather than a passive process; it constructs rather than records "reality."[19]

The book describes how we constantly misinterpret information—that is why different people reach different conclusions about reality. Fortunately, there are many ways to get around these biases and thus perceive the world through a wider and less distorted lens. Here are eight strategies that the CIA uses to teach its intelligence-gathering analysts to think more wisely and open-mindedly:

1. Become proficient in developing alternative points of view.
2. Do not assume that the other person will think or act like you.
3. Think backward. Instead of thinking about what might happen, put yourself into the future and try to explain how a potential situation could have occurred.
4. Imagine that the belief you are currently holding is wrong, and then develop a scenario to explain how that could be true. This helps you to see the limitations of your own beliefs.
5. Try out the other person's beliefs by actually acting out the role. This breaks you out of seeing the world through the habitual patterns of your own beliefs.
6. Play "devil's advocate" by taking the minority point of view. This helps you see how alternative assumptions make the world look different.
7. Brainstorm. A quantity of ideas leads to quality because the first ones that come to mind are those that reflect old beliefs. New ideas help you to break free of emotional blocks and social norms.
8. Interact with people of different backgrounds and beliefs.

On the surface, these suggestions seem easy, but they're not. Take, for example, steps 4 and 5, and apply them to your religious beliefs. Can you even imagine that your beliefs could be wrong? If so, can you conceive of how you could have been mistaken? What if there wasn't a God—how would your world be different? And if you're an atheist, and you discovered that God existed, how might that change your life? How many Democrats and Republicans have ever taken the time to see the world through the opposing political lens, and if they did, would they cooperate more readily to attain common ideals and goals? And how many parents, when they're angry at their child's behavior, take a moment to understand their own actions from the child's point of view? Wouldn't that be a better way to establish a dialogue?

Obviously, I have my own biases here, for I believe that the world would be safer if we all took time to see the universe through the eyes of as many people as possible. When an atheist can acknowledge the internal joy that a fundamentalist feels when contemplating the Bible, and when a fundamentalist can appreciate the atheist who is fighting for humanitarian ideals, there will be less hostility in the world. Becoming a better believer requires that you temporarily suspend your innermost beliefs. This takes courage.

Ask Questions and Double-Check Supposed Facts

If you want to be a better believer, ask lots of questions. Be curious and don't settle for superficial facts. Look closer, dig deeper, and investigate the source. Learn how to tell the difference between a personal opinion and actual data, and be open to modifying your beliefs. Then ask more questions, for questions help to expand your perceptions of the world. Approach your questioning with enthusiasm for finding truth rather than a desire to denigrate and tear down other people's beliefs. And most important, as I have emphasized throughout this book, keep in mind that we can never know for certain the accuracy of any beliefs, even those we hold most strongly.

Now, if you want an easy and enjoyable way to sharpen your critical thinking, pick up a copy of Michael Crichton's novel *State of Fear.*[20] By the time you're halfway through, you won't know whether to believe or disbelieve anything you've read about global warming, or even about secondhand smoke. You probably won't like what you read; but if you take the time to double-check it, you'll find that he makes a serious and valid point: even if the entire scientific community agrees on an issue, and receives the endorsement of environmental groups and governments, this does not guarantee that the consensus is correct.

With regard to science and medicine, it pays to be cautious about the information you read, especially when it is reported in newspapers, magazines, and the other popular media, because these sources often interpret data in ways that leave out essential information. I have found news stories that inadvertently turned information around so that the account contradicted what the original research said. This seems to happen frequently to studies on religion and spirituality, especially when statistical measures are used. So if you really want to know what a researcher said, you'd have to review the original report. That takes time, which is a precious commodity these days.

Many news stories also report preliminary findings that are often not supported when a study is finally completed. But reporters rarely cite the later versions, because the issue is no longer considered "news." And then there's the question of fraud. Since the 1980s, an increasing number of researchers at universities have fabricated essential components of their work. A recent scandal involved the procedures of Dr. Hwang Woo Suk of Seoul National University, who supposedly created the world's first cloned human embryo and Snuppy, the world's first cloned dog. But in December 2005, following allegations from his coworkers that he altered or fabricated some of his research, Suk publicly apologized and resigned from his university post.[21] With the number of international journals growing every year, there will be more inaccuracies and oversights, and occasionally outright fraud. That said, I think most researchers genuinely try to perform good research and interpret their results in an appropriate way. But even the most objective scientists still have their own biases.

Finally, there's the issue of reliability in popular books, especially those that report new advances in health. With the overwhelming amount of competing and conflicting research, it is easy for an author to selectively choose those studies that support the premise of the book, overlooking relevant contradictory information. But the author is not solely to blame, because what makes a book popular is the number of readers who purchase it, and most buyers want a book to be simple and fun to read. I think the best advice was given by Mark Twain, 100 years ago: "Be careful about reading health books. You may die of a misprint."

Surf for Knowledge

The Internet is an excellent place to begin a search for knowledge, but you have to know where to look and how to evaluate the websites where information is found. Keep in mind that most sites will include only material that promotes their own beliefs, so you'll have to surf elsewhere to hear the other side's point of view. If you don't, you won't be able to develop a comprehensive overview of the issue being discussed.

Let's say that you want to weigh the strengths and weaknesses of various evolutionary theories. You could go to a site that promotes intelligent design and view one set of opinions, or you could go to an academic site that would offer a different set of arguments. Both sites, however, will present only thumbnail sketches of their own views while giving you the mistaken impression that you've captured the whole perspective. Understanding something as complex as evolutionary biology takes years of reading and research. Of course, most of us don't take the time to do this, and so we rely on others to summarize the information for us. But the moment we do this, we become vulnerable to many of the cognitive biases I listed above. You'll be influenced unconsciously by authoritarian and publication biases, confirmation and self-serving biases, in-group and out-group biases, consensus and bandwagon biases, expectation and persuasion biases, and so on.

Then there's the problem of the information itself, which is just too much to absorb. The brain deals with excess neural information by ignoring it, and most people do the same when they walk into a bookstore or library. The information is there, but we're going to look only at what interests us most. Even an expert can barely stay abreast of his or her own field. In the year 2000, for instance, nearly 1 million articles were published in scientific journals, taking up enough paper to stretch from San Francisco to Tokyo. Add to these the 200,000 new books being published each year, and you'll understand why Stephen Hawking said, "If you stack all the new books being published next to each other, you would have to move at ninety miles per hour just to keep up with the end of the list."[22] You certainly wouldn't have any time to read.*

In the scientific community, we try to overcome this problem by using search engines to summarize and categorize information in ways that are easier to view. For example, if you wanted to evaluate the relationship between religion and health, you could go to the National Library of Medicine online—affectionately called PubMed—where you'll find one of the world's largest databases covering medical and psychological research. It won't include everything, for it does not track many small journals, but it will give you a good idea of the research being conducted. Of course, you should keep in mind the publishing bias I mentioned above, since journals are disinclined to publish studies with negative results. Most studies are listed with a one-paragraph abstract of the researcher's strategy, findings, and conclusion; but unless you read the full article (usually for a fee), you often will not know the strengths and weaknesses of the study. Abstracts sometimes suggest positive findings, but I have often come across articles in which the author states, at the end of the article, that the positive findings were statistically insignificant. It's also not uncommon to discover that the researchers based some of their own work on other studies that were seriously flawed, and this would obviously compromise the validity of present work.

* Actually, I think Hawking erred. His calculation was based on stacking 200,000 books next to each other. If each book were one foot long and the books were laid end to end, which is being generous, that adds up to less than thirty-eight miles. A brisk stroll at four miles per hour should cover this distance in half a day, and you'd have to read only 550 books a day to stay current. Even a great physicist can make mistakes.

Recognize the Limitations of Belief

The next step in becoming a better believer is to recognize that the maps we build can only approximate the truth about the world. Thus there will always be a fundamental gap between our knowledge, our beliefs, and reality.

For some people this is an uncomfortable thought, but for others it is uplifting. I personally chose to focus on brain research because we know so little about how the brain really works, and I am certain that a decade from now we will probably have frameworks entirely different from those we are using today. In fact, medical research advances so fast that we continually change the way we treat patients. Often this is felt as a burden, especially when a popular drug or procedure is shown to be dangerous for some patients. We all like to believe that the medications we take are safe, but in reality, physicians and drug researchers barely understand the mechanisms behind many of the drugs we use. For example, Prozac is one of the most widely used antidepressants, yet we don't know exactly why it works.

This lack of knowledge is somewhat scary, but it is also the engine that drives our scientific work. So the medical community generally takes great pride in knowing that it really doesn't know; this teaches us humility and strengthens our resolve to find better ways to improve our health.

Being uncertain has its advantages, but it means that we can never capture the truth. Socrates embraced such a philosophy by questioning everything and everyone in his search for wisdom, and he would often feign ignorance to bring out other people's beliefs. Then he would point out the errors in their logic. The power of his arguments terrified the Greek politicians because it challenged their moral beliefs, and so they put Socrates to death. Even then he declared, "I am only too conscious that I have no claim to wisdom, great or small."[23] For Socrates, the power was in the question, not the belief.

Develop Flexible Beliefs

By questioning our own beliefs and recognizing their limits, we open ourselves to exploring different systems of belief. And as we learn from others, the boundaries that normally separate us begin to dissolve.

My late colleague Gene d'Aquili was well known for his stimulating dinner parties. One evening, he brought together twelve distinguished individuals, including a judge on the state supreme court, a Nobel laureate, a rabbi, and a priest. The banquet began with a hearty soup, but unfortunately it contained chopped bacon. This presented a problem for the rabbi, who kept kosher, as well as for the priest, because this was the middle of Lent. Neither could consume the soup, since their religious tenets forbade it.

Suddenly Gene stood up and waved his hand over the soup, chanting in Latin, "*In conspectu Dei, omniumque in coelo et in terra habitantium, solemniter declaramus quod omnia super hanc lanceam posita, nunc pisces sunt.*" Then he translated what he had said: "In the sight of God and of all who dwell in the heaven and on the earth, we solemnly declare that everything placed upon this plate is now fish!" Jesus had converted water to wine, and Gene now claimed to have transformed bacon into fish, so you can imagine the consternation he caused.

The argument was simple: if the contents were truly fish, then everyone could eat the soup without transgressing any religious belief. Now some people might have considered Gene's act insensitive and sacrilegious, but the Jesuit priest smiled and laughed. "In the Judeo-Christian tradition," he announced, "it has been argued that the rule of the house can supersede religious doctrine if done properly and with respect for the traditions." The rabbi agreed, and so everyone proceeded to eat the soup. Being flexible saved the evening, but in the world at large, inflexible beliefs can ruin a person's life.

As we explore different systems of belief, we may find ourselves at a juncture. In childhood, most of our beliefs are given to us; and we do not question them deeply, because our brains are not mature enough to do so. But in adolescence, we do have the capacity to think for ourselves. Unfortunately, the power of group consensus inhibits individuality, because the group demands that its members conform to its beliefs. If, in the process of becoming a better believer, we challenge these moral tenets, we are faced with new and difficult challenges. For example, we may suddenly find ourselves outside the group, and this is a lonely path. Nor is it necessarily wise, for living in isolation can lead to physical and emotional stress. So we compromise, positioning ourselves between the two opposing poles of social conformity and independent autonomy.

Having flexible beliefs allows us to stand between opinions, and I believe that in this state, more meaningful dialogues emerge. Imagine being in a room with fifty people, where everyone is open to everyone else's ideas. It actually happens every day, in boardrooms across America. Creative people sit down with each other for the sole purpose of dreaming up new ideas. What eventually emerges is a hybrid invention, the merging of a new belief with an old behavior or technology. And it happens in scientific research as well, since most studies have two or more authors.

Avoid Stress—It Undermines Healthy Beliefs

In considering different beliefs, it is also important to assess which beliefs have a destructive effect on oneself or others. One way to do this is to evaluate the stress that certain beliefs place on your body and your mind. Stress itself, whether physical or psychological, plays a major role in every aspect of life. A little bit is good for you because it stimulates physical and psychological growth. But too much stress has the opposite effect, for the hormones that are released can permanently injure your brain.[24]

As I mentioned earlier, one of the first structures to be hit by stress is the hippocampus, which regulates emotion, memory, learning, and personality—those mechanisms that are essential in forming and maintaining healthy behaviors and beliefs.[25] Stress also influences other body systems that control growth, sexual and reproductive function, heart and respiratory rate, blood pressure, and digestion.[26]

In fact, stress can undermine every aspect of our cognitive and emotional stability, and the longer we remain stressed, the more it alters our perceptions of reality. In this way, stress causes a highly anxious or depressed individual to believe that the world really is a dangerous and unhappy place to live.

Ongoing stress gives rise to beliefs that one is helpless, hopeless, and emotionally crippled, and these thoughts trigger the release of even more stress hormones. Serotonin and dopamine levels also drop, weakening our ability to be rational or find creative solutions to our problems. This, in turn, disrupts frontal-lobe activity, making it all the more difficult to control feelings of depression, anxiety, and rage. Empathy and compassion—for others and for oneself—rapidly decline to the point where medical or psychological intervention is needed.

Stress, like pain, comes in many forms and has a cumulative effect. Multiple time commitments, financial problems, marital conflicts, lack of sleep, poor nutrition, illness, aging, moving, and dealing with illness or death are all stressful for the body and mind. Even exciting and happy events like marriage, childbirth, or getting a new job can add stress to one's life.[27] Other stressors include bad weather, environmental noise, traffic, and overcrowding. However, if you can maintain a positive attitude and receive adequate support from family, friends, and social groups, then you can cope with stress more easily.

Interrupt Negative Beliefs

Certain thoughts and beliefs also cause stress. These include, to name just a few: having unrealistic expectations for ourselves and others, having excessive guilt, being overly competitive, procrastinating, and jumping to conclusions too quickly. To this list, Albert Ellis, one of the founders of cognitive therapy, adds the following: illogical overgeneralizing ("I'm a bad person when I do bad things"), "awfulizing" ("This is terrible . . ."), "musturbating" ("I must . . . ," "I have to . . . ," "I should . . ."), and dividing everything into all-or-none categories ("I always . . . ," "I never . . . ,").[28] These forms of destructive thinking are based on inaccurate, inadequate, and pessimistic belief systems that undermine health.

In a study that included 12,000 women, researchers were able to determine that those who had negative beliefs about themselves were at more risk of developing depression, and that those who had the fewest negative self-beliefs were the least likely to become depressed. Obviously, optimistic beliefs are healthier, but they require that we ignore, to various degrees, the negative aspects of life. This form of "selective perception bias" is part of the brain's neurological mechanism for maintaining optimal health.[29]

Diffuse Emotions That Generate Destructive Beliefs

In a recent overview of the neural mechanisms of mood disorders, research found that emotional disturbances cause abnormalities to develop in many structures and functions of the brain. The disturbances include irregular cerebral blood flow and glucose metabolism, interruptions of memory storage and retrieval, impaired social empathy, and decreased volume and atrophy of various parts of the brain.[30] In addition, postmortem studies have linked emotional disturbances to abnormal reductions in glial cells, neuron size, and synaptic proteins. Simply put, anxiety, depression, and mania all wreak havoc on the brain. An emotionally distraught individual takes things out of context and reacts inappropriately, not because he or she is psychologically unsophisticated but because various parts of the brain, like the hippocampus and thalamus, are functioning poorly.

Negative emotions generate negative thoughts and beliefs, and together they disrupt other neural circuits in the brain. The good news is this: there is strong evidence that you can consciously interrupt these destructive neural processes by changing the way you think, be it through meditation, prayer, or cognitive-therapeutic intervention. At the University of Montreal, researchers found that human subjects could voluntarily alter their mental processes and influence the electrochemical dynamics of a neural circuit that promotes emotional self-regulation. Such a circuit, the researchers said, "may implement one of the most remarkable human faculties that has emerged in the course of human evolution." They also emphasized that a failure to control negative emotional responses could have "disastrous psychological and social consequences."[31]

In neurological studies conducted at the University of Wisconsin, Richard Davidson found that people who dwelt on distressing episodes in their lives had markedly lower antibody levels after they were given an influenza vaccination, whereas people who spent time recalling happy memories developed high antibody levels. In simple terms, negative thoughts and emotions weaken our ability to fight off disease.

Davidson added that when we are faced with uncertainty, the brain tends to engage a "negativity bias."[32] This suggests that optimism may be a learned behavior, dependent on what we believe and think about the world. Since the frontal lobes play an essential role in this process, they may help to explain the beneficial effects associated with various forms of meditation, prayer, guided imagery, and relaxation. Such programs can enable an individual to cope better with pain or discomfort, thus reducing stress, anxiety, and depression.[33]

I have recently been conducting a brain-scan study of a vigorous form of yoga known as Iyengar. We performed SPECT scans before and after a twelve-week training program, and the results, although preliminary, suggest that the practice of yoga may affect the brain in ways that may also alleviate depression and anxiety, possibly through a mechanism that optimizes the ability of the frontal lobes to regulate emotional responses. The results from our imaging study also suggest that the brain becomes better at entering states of meditation and relaxation the more often these are practiced. However, we do not know how much difference is brought about through longer periods of practice.

The Placebo Effect

I want to touch briefly on the placebo effect because it lends substantial support to the evidence that optimistic beliefs promote physiological health. Today, the question is not whether placebos work, but how they work.[34] Depending on the condition being treated, the effectiveness of placebo treatments can range from 0 percent to 100 percent.[35] Pain is the condition most amenable to placebo treatments; this suggests that pain regulation is closely tied to the same neural mechanisms that control conscious awareness and memory. Suggestibility, expectations, conditioning, emotions, and desires also play essential roles in explaining the analgesic effects of a placebo.[36] Herbert Benson of Harvard calls the effect "remembered wellness."[37]

Another intriguing finding is that placebos have different degrees of effectiveness in different cultures and countries.[38] This suggests that the physician's enthusiasm and faith in the treatment dramatically affect the outcome for the patient's health. It provides the clearest reason why Mr. Wright was able to fight off his cancer, at least for a period of months.*

Likewise, negative beliefs can cause a patient to get worse. This is known as the "nocebo" effect and can be brought on by a patient's false expectations, negative conditioning (e.g., the belief that something bad is going to happen), depressive thinking, and anxiety-related thoughts.[39] Mr. Wright was strongly affected when the American Medical Association announced that the drug he was taking was useless. In a matter of days, his tumors returned and he died.

In illnesses involving depression and anxiety, the placebo effect may account for a success rate of 25 to 35 percent. Antidepressant drugs have a success rate of 35 to 45 percent according to statistics published by the drug companies themselves. This suggests to me that a person's optimistic belief about getting well is largely responsible for the alleviation of depressive symptoms. On the other hand, the fact that nearly two-thirds of depressed individuals do not get better may be related to deeply embedded pessimistic beliefs.

From a physiological perspective, negative beliefs are probably associated with a complex array of brain functions, and thus a drug affecting only one function of the brain is not likely to cure depression. According to David Morris of the University of Virginia, placebos "place belief and meaning at the center of the therapeutic encounter," and "positive beliefs in the efficacy of medication or

*See the beginning of Chapter 1 for the story of Dr. Klopfer and Mr. Wright.

treatment are necessary to underwrite a placebo effect, while disbelief actively subverts it."[40]

In one study, asthmatics were exposed to water vapor but were told that they were inhaling irritants or allergens. Nearly half had an allergic reaction, but when they were given the same saline solution and told that it was a therapeutic treatment, their symptoms were relieved.[41] In other words, positive beliefs had the power to heal, whereas negative beliefs had the power to injure. This framework can be applied to spiritual beliefs as well. One can even speculate that those who do not inherit a bias toward optimistic beliefs are less likely to survive and pass on their genes to others.[42]

The power of placebo goes a long way to explain a variety of health claims made by alternative medicine and psychotherapy, for the high degree of success achieved may be attributed to the optimistic beliefs of the patient and the doctor. This holds true for traditional Western medicine as well, for patients are more likely to have a better outcome from medical or surgical interventions when they have confidence in their doctor, and in themselves.

Living with Our Beliefs

Placebo studies, prayer studies, and states of consciousness research bring to light an important dimension of the human spirit that is sometimes overlooked by science: our brain does not need absolute proof about anything. Instead, it seeks solutions for problems in a variety of creative ways. And this is an important point to keep in mind when we examine our deepest beliefs: they don't necessarily have to be accurate; they only have to help us survive. But they can also do much more, providing us with a sense of hope and optimism about ourselves and the world.

Becoming a better believer is a difficult task to undertake, for rewiring the brain requires patience and time. But if we succeed, to some small degree, then we will be better able to recognize our limitations, as well as our strengths. For this reason, I hold the deepest respect for those people who have had the courage to question and challenge their beliefs, for these are the individuals who have enriched our world through their creativity and willingness to grow.

Epilogue

Life, the Universe, and Our "Ultimate" Beliefs

A LONG, LONG TIME AGO—ACCORDING TO THE *HITCH-hiker's Guide to the Galaxy*—the second greatest computer in all creation was built to ponder the ultimate answer to life, the universe, and everything.* The computer was named Deep Thought, but the idea that an ultimate answer could be found greatly upset the philosophers of the time, for they feared that Deep Thought would put them out of business. The computer briefly commiserated with the academics, but told them they needn't worry too much because it would take 7.5 million years to solve the problem, and during that time, they'd have plenty of business trying to guess what the answer would be. So, 7.5 million years later, a great gathering came to hear Deep Thought's pronouncement. Yes, it had found the answer to life, the universe, and everything. But it was also convinced that no one was going to be pleased. Why? Because the answer was . . . forty-two.

"Forty-two!" yelled Loonquawl [a descendent of the person who built Deep Thought]. "Is that all you've got to show for seven and a half million years' work?"

* The greatest computer was . . . Well, I don't want to give away the ending, so you'll have to read the book. Better yet, get a recording of the original BBC broadcast. It should keep you entertained, no matter what planet you happen to be from.

"I checked it very thoroughly," said the computer, "and that quite definitely is the answer. I think the problem, to be quite honest with you, is that you've never actually known what the question is."

Perhaps Deep Thought was right: in our search for ultimate truth—be it about love or politics or God—we may be asking the wrong questions.

The Ultimate Question

Growing up, I was always inquisitive, and my father used to tease me by saying that I asked more questions than a congressman interviewing a Supreme Court nominee. I wondered about how everything worked—automobiles, planes, and of course that fantastic machine, the brain. From an early age, I tried to grasp the meaning and purpose of life, and would argue with my father into the wee hours of the night. But all my questioning never led me to the answer to what I now consider the ultimate question, which first arose when I attended summer camp in the Pocono Mountains of Pennsylvania. I was twelve years old at the time.

It had been a very hot afternoon, filled with activities, and we were exhausted by the end of the day. We'd had a particularly brutal baseball game, and were caked with dirt and sweat. The oppressive heat persisted into the night, so no one was able to fall asleep. Our minds wandered in all sorts of odd directions, and my friend Jim came up with the idea that it wasn't really hot; we only thought it was. "You're right!" I said. "It's only hot because we believe it's hot."

"Maybe everything is a fantasy," Jim speculated, "and maybe things are only real because we think they're real." I agreed, and in that moment, I raised the ultimate question: "How do we know if anything is real?" After all, I thought, everything may just be a dream. Such questions give adolescents headaches, so we promptly fell asleep. But I began to dream about baseball games in the heat, and in the dream, I wondered whether I was really asleep.

The next morning was blisteringly hot, and although we had forgotten our conversation of the previous night, we didn't seem to be bothered by the heat. In fact, we had more fun than ever.

The ultimate question returned when I took a college course in

Buddhism and western thought. My professor, Masao Abë, was about seventy years old and had traveled all the way from Japan to teach the course. I was intrigued by the concepts of this great tradition, especially those that pertained to the human experience of reality. In the Buddhist model, reality itself, as we perceive it, is merely an illusion of the mind.

One day I asked Dr. Abë to discuss this issue further. When I shared with him my ultimate question—"How do we know what's real?"—he smiled and said that there was no way to escape the illusions generated by the human mind. Even meditation, which could show you that perception was an illusion, could not do anything more. I was very disappointed, and said that there must be a way to get around the problem. He chuckled, then told me in a slightly condescending voice to come back to him when I had the problem solved.

By the end of the semester, my ultimate question receded into the hidden corners of my mind, but it didn't disappear for long. One day, while attending a program at the Dolphin Research Center in Florida, I was asked to participate in an experiment comparing human and dolphin brains. The test was simple. First, you listened to a tone, and then a second tone was played. If you thought it was the same as the original tone, you pushed one button, but if it sounded different, you'd push another button. If you were uncertain, you'd push a button designated "I don't know." I was rewarded points for my correct answers, and the dolphin was rewarded fish.

After completing the test, I was somewhat dismayed to discover that my degree of accuracy was almost identical to the dolphin's scores. Both of us seemed to know what was and wasn't real, and as I walked along the beach reflecting on this experience, I began to wonder what dolphins actually knew. Would they see the same moon rise and be equally touched by its beauty? The test suggested that they perceive some aspects of reality in much the same way that we do, and that they may have similar beliefs about the world. But perhaps they saw reality differently, for after all, they have a different type of brain. This raised the question whether a dolphin's experience of reality would be more or less accurate than my own. At the time, I didn't have a clue how to address such issues, but I believe that my "ultimate question" encouraged me to delve more deeply into the mechanisms of the human brain.

The ultimate question fully entered my life during the writing and publishing of my previous book, *Why God Won't Go Away.* The research I conducted strongly supported the notion that our perceptions remain locked within the human brain; and given the research that has followed, which I have presented in the current book, it seems that our perceptions are further restricted by the beliefs we consciously construct about the world. So far, I haven't been able to disprove my Buddhist professor's opinion, for it appears that our individual reality is solely guided by a combination of our sensory perceptions and our beliefs. But this hasn't stopped me from wondering, for now I ponder whether some beliefs are more accurate than others. If so, then those beliefs might at least bring us closer to knowing what is ultimately real.

Does Anything Exist Beyond Our Beliefs?

All sorts of individuals—physicists, cosmologists, biologists, geologists, theologians, and even some children—are striving for proof of what is fundamentally real. For some people, science is their savior. For others, it is philosophy or religion. For me, all are essential to explore if you're searching for the meaning of life.

In most schools of philosophy, proof is inextricably bound to logic, reason, and personal experience, but as I have demonstrated throughout this book, personal experience is subject to numerous perceptual, emotional, and cognitive distortions that occur at every stage of neural processing. What is finally summoned forth into consciousness turns out to be a very limited and subjective view of the world.

Science tries systematically to utilize subjective experience to measure objective reality, but even scientific views of reality differ. Every hypothesis finds dissenting views, so scientists themselves are challenged to choose who or what to believe. Furthermore, a scientist's belief system can influence the outcome of a study as much as a theologian's belief system can influence his or her perception of the world.

Science, like religion, can also be taken to a dogmatic level. This is known as "scientism," the belief that science will someday tell us everything we need to know about the universe. Scientism may turn out to be true, but the scientific perspective still remains bound by the limitations of our own consciousness; that is why we might never be able to provide a complete understanding about the world. Religion, of course, faces the same problem.

Still, scientists do have moments of insight—like the epiphanies of Archimedes, Newton, and Einstein—that seem to surpass everyday cognition and yield unexpected truths. These "aha" experiences are, in essence, altered states of consciousness, and they can be triggered by dreams, intuition, imaginative play, or various forms of meditation and prayer. It is my belief that when we are in such states, we can glimpse a different, and perhaps more accurate, view of reality. Whether or not this actually happens is unknown, but I believe that brain-scan research will shed significant light on where such insights come from.

To study spiritual experience scientifically, one needs a test subject and an observer—someone to make the measurements and interpret the results. Unfortunately, the observer can never peer into the mind of the subject to determine exactly what he or she is experiencing, or when. This has always been a problem in studying meditation and prayer. We can't tap the subjects on the shoulder and ask them where they were in their meditation, for the interruption would alter the very state we were trying to capture. This is the brain's version of Heisenberg's uncertainty principle—as we measure one quantity more precisely, the precision of the other quantity is less. There are equivalent principles in other fields of knowledge. Anthropologists, for example, know that their presence alters the culture they're observing. Likewise, mathematicians must grapple with Godel's incompleteness theorem, which suggests that there may always be assumptions that are impossible to prove. Even the machines neurologists use to measure brain activity have built-in limitations to their accuracy; consequently, medicine is as much an art as a science.

In my investigation of mystical experiences, my subjects often believe that they perceive a higher or deeper layer of reality to which they feel connected. Furthermore, they often feel that the experience is profoundly real. From my research with Buddhist meditators, Franciscan nuns, and Pentecostalists, I have come to believe that the more one focuses on a certain belief, the more real it may ultimately feel, and that this sense of realness is based on the stimulation of specific neural circuits in the brain. Emotions and sensory stimulation also accentuate this sense of realness; that is why many religious rituals incorporate dramatic movement, sound, and light. What the research has found so far is that your perceptions of reality and your beliefs are inextricably intertwined.

Determining What Is Real

Your sense of reality depends primarily on three criteria: the subjective vividness of an experience, the continuity and duration of the experience through time and space, and the consensus of others on what is considered real. For me, all three criteria can be reduced to the first one—the vivid sense of any experience ultimately defines what is real and what is not. For example, your perceptions of time, space, and duration are subjectively structured experiences created by the brain; if you alter the brain's functioning for any reason, a significant perceptual shift will occur in all these dimensions. Thus, time, space, duration, and even your sense of self are subjectively relative experiences that can change from moment to moment, depending on the neural activity of the brain.

The third criterion for establishing what reality is—interpersonal validation—is also subjective. The fact that someone agrees with you does not necessarily make your opinion true; and yet, as far as the brain is concerned, validation from others is essential for determining truth. Even when people see totally different objects, research consistently demonstrates that peer-group pressure unconsciously forces us to align our opinions.

Therefore, our subjective experience becomes the sole arbiter of what we consider real. We reach out and touch a table. It feels real, and because it feels real, we believe that it is real. When we dream, our dreams too seem real, in part because various cognitive systems have been suspended. But when we awake, our senses reestablish themselves, and so we interpret our experience as "less real" according to our preestablished beliefs.

The same thing happens when we have a spiritual experience: we suspend certain sensory processes and beliefs and accentuate others in ways that allow the experience to feel real, but this condition lasts only while we are having the experience. When it ends, we return to baseline awareness and categorize the experience in one of several ways. For example, we might argue that the experience was a distortion or hallucination, or we can argue that what we experienced was superreal—more real than our everyday reality. In either case, we usually base our decision on previously held beliefs.

What if it were possible to suspend all systems of belief? In Buddhism, this is known as pure consciousness, a state in which one is awake and aware, yet no thoughts are being consciously processed or perceived. The rare individual who reaches such a state must be interrupted by someone else in order to eat or sleep. What would be the benefit of such a practice? According to the Buddhists, this state brings a profound sense of peacefulness that is free from the sufferings caused by desires, worries, and fears. The present is all that exists. When these people come back to everyday reality—when they reengage in the world—they appear to have more wisdom and compassion than other individuals; and when you are in their presence, they seem to radiate a profound sense of peace. What interests me about these states is that the person may have found a way to step outside of his or her system of belief and thus experience reality more open-mindedly. If we can find such gifted individuals, we now have the technology to peer into their brains to help us understand the reality they perceive.

What do I expect to find? Hopefully, I might stumble on the connection between the "everyday" reality created by the brain and the fundamental reality that links us to life, the universe, and everything. And the more we push science and spirituality to their limits, the better chance we have to answer the "ultimate" question.

In *Hitchhiker's Guide to the Galaxy,* the ultimate question turned out to be a line from Bob Dylan's *Blowin' in the Wind,* which asked how many paths must a man have to walk before he could be called a man. It's not a bad question, and I, too, have often paraphrased

Dylan, wondering how many paths we must walk before we discover the truth, but I doubt that the answer will be forty-two.

In the end, we must always return to our beliefs. From the mundane to the mystical, they inform us about reality and they shape our future lives. And if the ultimate reality remains a mystery, so much the better, for it is the questions that give us meaning, that drive us forward and fill us with transcendent awe.

At least, that is what I believe.

Endnotes

Chapter 1: The Power of Belief (Pages 3–15)

1. B. Klopfer. 1957. Psychological variables in human cancer. *Journal of Projective Techniques* 21.

2. B. O'Regan and C. Hirshberg. 1993. *Spontaneous Remission: An Annotated Bibliography.* Petaluma, CA: Institute of Noetic Studies. (The entire book, along with a useful page of FAQs, can be found on the website for the Institute of Noetic Studies: http://www.noetic.org/research/sr/faqs.html.

3. M. Heim and R. Schwarz. 2000. Spontaneous remission of cancer: Epidemiological and psychosocial aspects. *Zeitschrift für Psychosomatische Medizin und Psychotherapie* 46(1):57–70.

4. R. J. Papac. 1998. Spontaneous regression of cancer: Possible mechanisms. *In Vivo* 12(6):571–578. See also J. Markowska and A. Markowska. 1998. Spontaneous tumor regression. *Ginekologia polska* 69(1):39–44. And see R. Schwarz and M. Heim. 2000. Psychosocial considerations about spontaneous remission of cancer. *Onkologie* 23(5):432–435.

5. K. Armstrong. 1997. *A History of God.* New York: Random House.

6. Harris Poll No. 60, October 16, 2003.

7. This passage by Chuang Tzu was translated for this volume by M. R. Waldman.

8. *Born to Believe* is the natural next step in a line of research and scholarship which began in 1975 when my late colleague Eugene d'Aquili and Charles Laughlin published "The biopsychological determinants of religious ritual behavior" in *Zygon, Journal of Religion and Science.* Their thesis was that all religious phenomena were associated with various neuropsychological processes within the human organism. The first ten years of this scholarly movement were difficult, and the attempt to integrate neuropsychology and theology was extremely controversial. In 1979, Eugene d'Aquili (with Charles Laughlin and John McManus) published *The Spectrum of Ritual* (Columbia University Press, 1979). The following articles were published by Dr. d'Aquili in *Zygon:* "The neurobiological bases of myth and concepts of deity" (1978), "Senses of reality in science and religion" (1982), and "Myth, ritual and the archetypal hypothesis: Does the dance generate the word?" (1986). Other early explorations of the neuropsychological nature of religious experience were made by Nobel laureate Roger Sperry, Colwyn Trevarthen, Solomon Katz, Herbert Benson, Victor Turner, Laurence McKinney, and James Ashbrook. Ashbrook first used the term "neurotheology" in an article published in *Zygon* in 1984 entitled "Neurotheology:

The working brain and the work of theology." Ashbrook broadly defined neurotheology in terms of the split-brain physiology being researched at that time, a speculative vision that was yet to be substantiated by research (a Medline search found forty-one academic articles published between 1973 and 1996 relating to meditation and the brain).

In 1991, I began working with Dr. d'Aquili and others, creating a series of theoretical articles that led to a number of brain imaging studies that we first presented in 1993 (for example: "Religious and mystical states: A neuropsychological model," published in *Zygon* in 1993, "The near death experience as archetype: A model for 'prepared' neurocognitive processes," published in *The Anthropology of Consciousness* in 1994, "The neuropsychological basis of religion: Or why God won't go away," published in *Zygon* in 1998, "The neurophysiological correlates of meditation: Implications for neuroimaging," published in the *Journal of the Indian Academy of Clinical Medicine* in 1998, and "The neural basis of the complex mental task of meditation: Neurotransmitter and neurochemical correlates," published in *Medical Hypotheses* in 2003. Over the past ten years, my colleagues and I continued our brain-imaging studies of various religious and spiritual practices, including the more recent ones that you will read about in this book. During this time, many new research articles, books, and conferences began to address the relationship between spirituality and the brain. In 1996, a consensus conference on spirituality and health, convened by the National Institute of Healthcare Research, laid out numerous ideas concerning future research in this area.

Many other researchers and authors have made significant contributions to the field. These include Herbert Benson (*Timeless Healing: The Power and Biology of Belief,* Scribner, 1996), James Austin (*Zen and the Brain,* MIT Press, 1998), Matthew Alper (*The "God" Part of the Brain,* Rogue Press, 1998), Michael Shermer (*How We Believe: Science, Skepticism, and the Search for God,* W. H. Freeman, 1999), Elio Frattaroli (*Healing the Soul in the Age of the Brain,* Viking, 2001), Pascal Boyer (*Religion Explained: The Evolutionary Origins of Religious Thought,* Basic Books, 2001), John Horgan (*Rational Mysticism: Dispatches from the Border Between Science and Spirituality,* Houghton Mifflin, 2003), Joseph Giovannoli (*The Biology of Belief,* Rosetta Press, 2001), Dean Hamer (*The God Gene: How Faith Is Hardwired into Our Genes,* Doubleday, 2004), and Bruce Lipton (*The Biology of Belief: Unleashing the Power of Consciousness, Matter and Miracles,* Mountain of Love, 2005). These individuals, along with many others, have helped to stimulate important dialogues and research in many interdisciplinary fields.

9. J. Narby. 2001. *Shamans Through Time.* New York: Tarcher/Putnam.

10. C. Sagan. 1996. *The Demon-Haunted World.* New York: Ballantine.

11. R. Ader and N. Cohen. 1982. Behaviorally conditioned immunosuppression and murine systemic lupus erythematosus. *Science* 215, 1534–1536.

12. T. D. Wager and J. B. Nitschke. 2005. Placebo effects in the brain: Linking mental and physiological processes. *Brain, Behavior, and Immunity* 19, 281–282.

13. J. M. Blom, L. Tamarkin, J. R. Shiber, and R. J. Nelson. 1995. Learned immunosuppression is associated with an increased risk of chemically-induced tumors. *Neuroimmunomodulation* 2(2):92–99.

14. A. Brannstrom and U. Dieckmann. 2005. Evolutionary dynamics of altruism and cheating among social amoebas. *Proceedings of Biological Sciences* 272(1572): 1609–1616.

15. B. Mangan. 2003. Volition and property dualism. *Journal of Consciousness Studies* 10, 29–34.

16. S. Schneider, J. Miller, E. Crist, and P. Boston (eds.). 2004. *Scientists Debate Gaia.* Cambridge, MA: MIT Press.

17. T. Sugimoto. 2002. Darwinian evolution does not rule out the Gaia hypothesis. *Journal of Theoretical Biology* 218(4):447–455.

18. C. A. Eastman. 1911. *The Soul of an Indian.* New York: Houghton Mifflin.

Chapter 2: A Mountain of Misperceptions (Pages 16–44)

1. See L. Baker. 1995. *Explaining Attitude: A Practical Approach to the Mind.* Cambridge Studies in Philosophy. London: Cambridge University Press.

2. O. Judson. 2003. *Dr. Tatiana's Sex Advice to All Creation.* New York: Owl.

3. J. Roughgarden. 2004. *Evolution's Rainbow: Diversity, Gender, and Sexuality in Nature and People.* Berkeley: University of California Press.

4. Merriam-Webster's Online Dictionary.

5. R. Carter. 2002. *Exploring Consciousness.* Berkeley: University of California Press.

6. D. Schacter and E. Scarry. 2000. *Memory, Brain, and Belief.* Cambridge, MA: Harvard University Press.

7. D. Dennett. 1991. *Consciousness Explained.* New York: Little, Brown.

8. W. Parrott. 2001. *Emotions in Social Psychology.* London: Psychology Press.

9. J. Panksepp. 1998. *Affective Neuroscience: The Foundations of Human and Animal Emotions.* Oxford: Oxford University Press.

10. O. Sacks. 1985. *The Man Who Mistook His Wife for a Hat.* New York: Summit.

11. S. Feferman et al. (eds.). 1986. *Kurt Gödel: Collected Works,* Vol. I: *Publications 1929-1936.* Oxford: Oxford University Press.

12. D. Hofstadter. 1979. *Gödel, Escher, Bach: An Eternal Golden Braid.* New York: Basic Books. In his 1951 Gibbs lecture, "Some basic theorems on the foundations of mathematics and their philosophical implications," Gödel also believed that his theorem could be applied to human intelligence, arguing that the human mind cannot formally prove its own consistency. Similar controversial arguments have been taken up by the philosopher Hilary Putnam and the mathematician Roger Penrose.

13. R. Sapolsky. 1998. *Why Zebras Don't Get Ulcers: An Updated Guide to Stress, Stress-Related Diseases, and Coping,* 2nd ed. New York: Freeman.

14. Dozens of books have been written debating Spinoza's concept of substance. For a comprehensive analysis of these arguments, see R. S. Woolhouse. 1993. *Descartes, Spinoza, Leibniz: The Concept of Substance in Seventeenth-Century Metaphysics.* Oxford: Routledge. See also E. Harris. 1995. *The Substance of Spinoza.* Atlantic Highlands, NJ: Humanities.

15. B. Spinoza. 1991. *The Ethics: Treatise on the Emendation of the Intellect and Selected Letters.* Indianapolis, IN: Hackett.

16. A. Damasio. 1994. *Descartes' Error: Emotion, Reason, and the Human Brain.* New York: Avon.

17. A. Damasio. 2000. Thinking about belief. In D. L. Schacter and E. Scarry (eds.), *Memory, Brain, and Belief.* Cambridge, MA: Harvard University Press.

18. Ibid.

19. A. Damasio. 2003. *Looking for Spinoza: Joy, Sorrow, and the Feeling Brain.* New York: Harcourt.

20. H. Eichenbaum and J. A. Bodkin. 2000. Belief and knowledge as forms of memory. In D. L. Schacter and E. Scarry (eds.), *Memory, Brain, and Belief.* Cambridge, MA: Harvard University Press.

21. D. Dennett. 1997. *Kinds of Minds: Towards an Understanding of Consciousness.* New York: Basic Books.

Chapter 3: Reality, Illusions, and the Aunt Who Cried Wolf (Pages 45–69)

1. R. M. Shepard. 1990. *Mind Sights: Original Visual Illusions, Ambiguities, and Other Anomalies, with a Commentary on the Play of Mind in Perception and Art.* New York: Freeman.

2. R. L. Gregory. Perceptual illusions and brain models. *Proceedings of the Royal Society of London, Series B* 171:179–296.

3. U. Hasson, T. Hendler, D. B. Bashat, and R. Malach. 2001. Vase or face? A neural correlate of shape-selective grouping processes in the human brain. *Journal of Cognitive Neuroscience* 13:744–753.

4. For a comprehensive analysis of numerous optical illusions, see Al Seckel's *Illusionworks, L.L.C.* website at http://psylux.psych.tu-dresden.de/i1/kaw/diverses%20 Material/ www.illusionworks.com/index.html.

5. This illusion was created by Ludimar Hermann in 1870.

6. The first illusions demonstrating the distortion of parallel lines were published by Ewald Herring in 1861. In 1979, Richard Gregory noted the same illusion on the black-and-white tiled wall of a British café, which is why it is sometimes called the "café wall" illusion. R. Gregory and P. Heard. 1979. Border locking and the café wall illusion. *Perception* 8:365–380. This version was released into public domain by Lupin at the English Wikipedia project.

7. A. Seckel, see note 4 above.

8. E. Sampaio, S. Maris., and P. Bach-y-Rita. 2001. Brain plasticity: "Visual" acuity of blind persons via the tongue. *Brain Research* 908:204–207.

9. P. Weis. 2001. *Science News* 160, 140.

10. J. Davidoff. 2001. Language and perceptual categorization. *Trends in Cognitive Sciences* 5:383–387.

11. T. Lamb and J. Bourriau. 1995. *Colour: Art and Science.* London: Cambridge University Press.

12. J. Neitz, J. Carroll, and M. Neitz. 2001. Almost reason enough for having eyes. *Optics and Photonics News,* January 2001.

13. D. Deutsch. 2003. *Phantom Words and Other Curiosities,* Philomel Records, compact disc album booklet remarks.

14. J. S. Bruner. 1973. *Beyond the Information Given: Studies in the Psychology of Knowing.* New York: Norton.

15. T. S. Andersen, K. Tiippana, and M. Sams. 2004. Factors influencing audiovisual fission and fusion illusions. *Cognitive Brain Research* 21:301–308.

16. L. Shams, Y. Kamitani, and S. Shimojo. 2002. Visual illusion induced by sound. *Brain Research: Cognitive Brain Research* 14:147–152.

17. J. Bruner and C. Feldman. 1996. Group narrative as a cultural context of autobiography. In *Remembering Our Past: Studies in Autobiographical Memory,* ed. D. C. Rubin. London: Cambridge University Press, 291–317.

18. V. S. Ramachandran. 1998. *Phantoms in the Brain.* New York: Morrow.

19. M. Zampini, V. Moro, and S. M. Aglioti. 2004. Illusory movements of the contralesional hand in patients with body image disorders. *Journal of Neurology, Neurosurgery, and Psychiatry* 75:1626–1628.

20. Ramachandran, *Phantoms in the Brain,* op. cit.

21. C. M. Fisher. 1999. Phantom erection after amputation of penis. *Canadian Journal of Neurological Sciences* 26:53–56.

22. C. Gardner-Thorpe and J. Pearn. 2004. The Cotard syndrome: Report of two patients. *European Journal of Neurology* 11:563–566.

23. American Psychiatric Association. *Diagnostic and Statistical Manual of Mental Disorders (DSM-IV).* 1994. Washington DC: American Psychiatric Association.

24. F. E. Roux, D. Ibarrola, Y. Lazorthes, and I. Berry. 2001. Virtual movements activate primary sensorimotor areas in amputees: Report of three cases. *Neurosurgery* 49:736–741.

25. Z. Seltzer, T. Wu, M. B. Max, and S. R. Diehl. 2001. Mapping a gene for neuropathic pain-related behavior following peripheral neurectomy in the mouse. *Pain* 93:101–106.

26. D. J. Simons and C. F. Chabris. 1999. Gorillas in our midst: Sustained inattentional blindness for dynamic events. *Perception* 28:1059–1074.

27. Other interactive demonstrations of perceptual blindness can be found at http://rlandman.sdf-eu.org/change_blindness.htm (prepared by Rogier Landman, Ph.D., at the Laboratory of Neuropsychology, National Institutes of Mental Health) and at http://nivea.psycho.univ-paris5.fr/Mudsplash/Nature_Supp_Inf/Movies/Movie_List.html (prepared by J. Kevin O'Regan, Ronald A. Rensink, and James J. Clark at the Laboratoire de Psychologie Expérimentale, CNRS, Université René Descartes).

28. A. L. Barabasi. 2002. *Linked.* New York: Plume/Penguin.

29. D. J. Simons, S. R. Mitroff, and S. L. Franconeri. 2003. Scene perception: What we can learn from visual integration and change detection. In M. Peterson and G. Rhodes (eds.), *Perception of Faces, Objects, and Scenes: Analytic and Holistic Processes.* New York: Oxford University Press.

30. B. Kolb. 2004. Mechanisms of cortical plasticity after neuronal injury. In J. Ponsford (ed.), *Cognitive and Behavioral Rehabilitation.* New York: Guilford.

31. R. F. Anda, V. J. Felitti, J. D. Bremner, et al. 2005. The enduring effects of abuse and related adverse experiences in childhood: A convergence of evidence from neurobiology and epidemiology. *European Archives of Psychiatry and Clinical Neuroscience.* (Electronic publication November 29, preceding print.) See also M. H. Teicher, N. L. Dumont, Y. Ito, et al. 2004. Childhood neglect is associated with reduced corpus callosum area. *Biological Psychiatry* 56(2):80–85. And see J. D. Bremner, M. Vythilingam, E. Vermetten, et al. 2003. MRI and PET study of deficits in hippocampal structure and function in women with childhood sexual abuse and posttraumatic stress disorder. *American Journal of Psychiatry* 160(5):924–932.

32. M. Gazzaniga and J. LeDoux. 1978. *The Integrated Mind.* New York: Plenum.

33. B. H. Lipton. 2005. *The Biology of Belief,* Santa Rosa, CA: Elite. See also http://www.brucelipton.com/biology.php.

34. See http://ebiomedia.com/gall/eyes/octopus-insect.html.

35. R. Carter. 2002. *Exploring Consciousness.* Berkeley: University of California Press.

Chapter 4: Santa Claus, Lucky Numbers, and the Magician in Our Brain (Pages 70–99)

1. A. Conan Doyle. *A Study in Scarlet.* 1888. London: Ward, Lock. (First published in 1887 as the main part of *Beeton's Christmas Annual.*)

2. S. P. Johnson, J. G. Bremner, A. Slater, et al. 2003. Infants' perception of object trajectories. *Child Development* 74(1):94–108.

3. Research concerning infants' expectations has been carried on for decades. Developmental psychologists refer to it as "violation of expectations." An excellent overview of the literature, controversies, and conflicting findings can be found in Y. Munakata. 2000. Challenges to the violation-of-expectation paradigm: Throwing the conceptual baby out with the perceptual processing bathwater? *Infancy* 1(4):471–477.

4. D. Burnham. 1993. Visual recognition of mother by young infants: Facilitation by speech. *Perception* 22(10):1133–1153.

5. R. N. Desjardins and J. F. Werker. 2004. Is the integration of heard and seen speech mandatory for infants? *Developmental Psychobiology* 45(4):187–203.

6. K. H. Onishi and R. Baillargeon. 2005. Do 15-month-old infants understand false beliefs? *Science.* 308(5719):255–258.

7. For an overview of studies showing cognitive biases, see T. Gilovich. 1991. *How We Know What Isn't So: The Fallibility of Human Reason in Everyday Life.* New York: Free Press.

8. J. Metcalfe. 1998. Cognitive optimism: Self-deception or memory-based processing heuristics? *Personality and Social Psychology Review* 2(2):100–110.

9. N. D. Weinstein, S. E. Marcus, and R. P. Moser. 2005. Smokers' unrealistic optimism about their risk. *Tobacco Control* 14(1):55–59.

10. D. Lovallo and D. Kahneman. 2003. Delusions of success: How optimism undermines executives' decisions. *Harvard Business Review* 81(7):56–63, 117.

11. R. Sapolsky. 2004. *Why Zebras Don't Get Ulcers,* 3rd ed. New York: Owl.

12. S. C. Segerstrom, S. E. Taylor, M. E. Kemeny, and J. L. Fahey. 1998. Optimism is associated with mood, coping, and immune change in response to stress. *Journal of Personality and Social Psychology* 74(6):1646–1655.

13. J. C. Lai, P. D. Evans, S. H. Ng, et al. 2005. Optimism, positive affectivity, and salivary cortisol. *British Journal of Health Psychology* 10(Part 4):467–484.

14. For a comprehensive description of cognitive functions relating to mysticism and spirituality, see A. B. Newberg, E. G. D'Aquili, and V. Rause. 2001. *Why God Won't Go Away: Brain Science and the Biology of Belief.* New York: Ballantine, 2001. See also E. G. D'Aquili and A. B. Newberg. 1999. *The Mystical Mind: Probing the Biology of Religious Experience.* Minneapolis, MN: Fortress.

15. G. Lakoff and R. E. Nuñez. 2001. *Where Mathematics Comes From: How the Embodied Mind Brings Mathematics into Being.* New York: Basic Books.

16. J. N. Wood and E. S. Spelke. 2005. Infants' enumeration of actions: Numerical discrimination and its signature limits. *Developmental Science* 8(2):173–181.

17. K. Devlin. 2001. *The Math Gene: How Mathematical Thinking Evolved and Why Numbers Are Like Gossip.* New York: Basic Books.

18. B. Butterworth. 2005. The development of arithmetical abilities. *Journal of Child Psychology and Psychiatry* 46(1):3–18.

19. F. Xu, E. S. Spelke, and S. Goddard. 2005. Number sense in human infants. *Developmental Science* 8(1):88–101.

20. K. McCrink and K. Wynn. 2004. Large-number addition and subtraction by 9-month-old infants. *Psychological Science* 15(11):776–781.

21. E. B. Eamonn, and N. R. Franks. 2000. Ants estimate area using Buffon's needle. *Proceedings of the Royal Society of London Bulletin* 267(1445):765–760.

22. J. L. Deneubourg, S. Goss, N. Franks, and J. M. Pasteels. 1989. The blind leading the blind: Modeling chemically mediated army ant raid patterns. *Journal of Insect Behavior* 2:719–725.

23. T. S. Collett and D. Waxman. 2005. Ant navigation: Reading geometrical signposts. *Current Biology* 15(5):R171–173.

24. J. Best. 2004. *More Damned Lies and Statistics.* Berkeley: University of California Press.

25. At present, the notion that there is an inherent logic in genetic encoding and neurological development is primarily a theoretical model. Theories of "neural logic" are being investigated in fields as diverse as genetic programming, natural language processing, human reasoning, computer engineering, mathematical logic, and the logic of physics. For further information and references, see *Journal of Applied Logic.*

26. J. G. Quirk, E. Likhtik, J. G. Pelletier, and D. Paré. 2003. Stimulation of medial prefrontal cortex decreases the responsiveness of central amygdala output neurons. *Journal of Neuroscience* 23(25):8800–8807. See also the following. M. A. Morgan, L. M. Romanski, and J. E. LeDoux. 1993. Extinction of emotional learning: contribution of medial prefrontal cortex. *Neuroscience Letters* 163:109–113. M. R. Milad and G. J. Quirk. 2002. Neurons in medial prefrontal cortex signal memory for fear extinction. *Nature* 420:70–74. A. Bechara, H. Damasio, A. R. Damasio, and G. P. Lee. 1999. Different contributions of the human amygdala and ventromedial prefrontal cortex to decision making. *Journal of Neuroscience* 19:5473–5481.

27. J. A. Cheyne, S. D. Rueffer, and I. R. Newby-Clark. 1999. Hypnagogic and hypnopompic hallucinations during sleep paralysis: Neurological and cultural construction of the nightmare. *Consciousness and Cognition* 8(3):319–337.

28. O. Blanke, T. Landis, L. Spinelli, and M. Seeck. 2004. Out-of-body experience and autoscopy of neurological origin. *Brain* 127(Part 2):243–258.

29. W. Hirstein. 2004. *Brain Fiction: Self-Deception and the Riddle of Confabulation.* Cambridge, MA: MIT Press.

30. R. L. Watson, S. F. Dowell, M. Jayaraman, et al. 1999. Antimicrobial use for pediatric upper respiratory infections: Reported practice, actual practice, and parent beliefs. *Pediatrics* 104(6):1251–1257.

31. J. Alcock. 1999. Alternative medicine and the psychology of belief. *Scientific Review of Alternative Medicine*, Fall-Winter.

32. C. A. Bagley, S. Ohara, H. C. Lawson, and F. A. Lenz. 2006. Psychophysics of CNS pain-related activity: Binary and analog channels and memory encoding. *Neuroscientist* 12(1):29–42. See also F. A. Lenz, S. Ohara, R. H. Gracely, et al. 2004. Pain encoding in the human forebrain: Binary and analog exteroceptive channels. *Journal of Neuroscience* 24(29):6540–6544.

33. A. Miller (ed.). 2004. *The Social Psychology of Good and Evil.* New York: Guilford.

34. H. Tajfel, M. C. Flament, M. Billig, and R. P. Bundy. 1971. Social categorization and intergroup behavior. *European Journal of Social Psychology* 1:149–178.

35. M. E. Wheeler and S. T. Fiske. 2005. Controlling racial prejudice: Social-cognitive goals affect amygdala and stereotype activation. *Psychological Science* 16(1):56–63. See also A. J. Hart, P. J. Whalen, L. M. Shin, et al. 2000. Differential response in the human amygdala to racial outgroup versus ingroup face stimuli. *Neuroreport* 11(11):2351–2355.

36. W. A. Cunningham, M. K. Johnson, C. L. Raye, et al. 2004. Separable neural components in the processing of black and white faces. *Psychological Science* 15(12):806–813.

37. E. Aronson. 2004. Reducing hostility and building compassion: Lessons from the jigsaw classroom. In A. Miller (ed.), *The Social Psychology of Good and Evil.* New York: Guilford.

38. George W. Bush, address to a joint session of Congress, September 20, 2001.

39. For a more detailed view of reductionist and holistic functions of the brain, along with references for each of the cognitive functions described in this chapter, see E. G. D'Aquili and A. B. Newberg. 1999. *The Mystical Mind: Probing the Biology of Religious Experience.* Minneapolis, MN: Fortress.

40. N. K. Logothetis. 2000. Object recognition: Holistic representations in the monkey brain. *Spatial Vision* 13(2–3):165–178.

41. S. E. Trehub and E. E. Hannon. 2006. Infant music perception: Domain-general or domain-specific mechanisms? *Cognition.* (Published electronically before print publication.)

42. I. Gauthier, T. Curran, K. M. Curby, and D. Collins. 2003. Perceptual interference supports a non-modular account of face processing. *Nature Neuroscience* 6(4):428–432. See also Y. Lerner, T. Hendler, D. Ben-Bashat, et al. 2001. A hierarchical axis of object processing stages in the human visual cortex. *Cerebral Cortex* 11(4):287–297.

43. M. Tovee. 1998. Is face processing special? *Neuron* 21:1239–1242. (Also, see sources in note 42 above.)

44. T. Sharon and J. D. Woolley. 2004. Do monsters dream? Young children's understanding of the fantasy/reality distinction. *British Journal of Developmental Psychology* 22, 293–310.

45. N. M. Prentice, M. Manosevitz, and L. Hubbs. 1978. Imaginary figures of early childhood: Santa Claus, Easter Bunny, and the Tooth Fairy. *American Journal of Orthopsychiatry* 48(4):618–628.

46. C. J. Boyatzis. 2005. Religious and spiritual development in childhood. In R. F. Paloutzian and C. L. Park (eds.), *Handbook of the Psychology of Religion and Spirituality.* New York: Guilford.

47. E. M. Evans. 2001. Cognitive and contextual factors in the emergence of diverse belief systems: Creation versus evolution. *Cognitive Psychology* 42(3):217–266.

48. C. L. Niebauer, S. D. Christman, S. A. Reid, and K. J. Garvey. 2004. Interhemispheric interaction and beliefs on our origin: Degree of handedness predicts beliefs in creationism versus evolution. *Laterality* 9(4):433–447.

49. M. White. 1998. *Isaac Newton: The Last Sorcerer.* New York: Perseus.

Chapter 5: Parents, Peas, and "Putty Tats" (Pages 103–131)

1. C. Westbury and D. Dennett. 2000. Mining the past to construct the future: Memory and belief as forms of knowledge. In D. Schacter and E. Scarry (eds.), *Memory, Brain, and Belief.* Cambridge, MA: Harvard University Press, p. 20.

2. Quotations published by the Gorilla Foundation, www.koko.org. See also F. Patterson and E. Linden. 1986. *The Education of Koko.* Austin, TX: Holt, Rinehart, and Winston.

3. J. Piaget. 1962. *Plays, Dreams, and Imitation in Childhood.* New York: Norton.

4. A. Fotopoulou, M. Solms, and O. Turnbull. 2004. Wishful reality distortions in confabulation: A case report. *Neuropsychologia* 42(6):727–744.

5. R. Hastie and R. M. Dawes. 2001. *Rational Choice in an Uncertain World: The Psychology of Judgement and Decision Making.* Thousand Oaks, CA: Sage.

6. E. F. Loftus and H. Hoffman. 1989. Misinformation and memory: The creation of new memories. *Journal of Experimental Psychology* 118(1):100–104.

7. E. F. Loftus. 2003. Make-believe memories. *American Psychologist* 58:867–873.

8. I. E. Hyman, T. H. Husband, and J. F. Billings. 1995. False memories of childhood experiences. *Applied Cognitive Psychology* 90:181–197.

9. E. F. Loftus and D. M. Bernstein. 2005. Rich false memories. In A. F. Healy (ed.), *Experimental Cognitive Psychology and Its Applications.* Washington, DC: American Psychological Association Press, pp. 101–113.

10. E. Krackow and S. J. Lynn. 2003. Is there touch in the game of Twister? The effects of innocuous touch and suggestive questions on children's eyewitness memory. *Law and Human Behavior* 27(6):589–604.

11. E. F. Loftus. 1995. Remembering dangerously. *Skeptical Inquirer.* 19(2):20.

12. E. F. Loftus. 1993. The reality of repressed memories. *American Psychologist* 48:518–537.

13. CNN. 1993. "Guilt by Memory." Broadcast on May 3.

14. H. I. Lief. 1999. Patients versus therapists: Legal actions over recovered memory therapy. *Psychiatric Times* 16(11).

15. E. Geraerts, E. Smeets, M. Jelicic, et al. 2005. Fantasy-proneness, but not self-reported trauma, is related to DRM performance of women reporting recovered memories of childhood sexual abuse. *Consciousness and Cognition* 14(3):602–612.

16. G. A. L. Mazzoni, P. Lombardo, S. Malvagia, and E. F. Loftus. 1999. Dream interpretation and false beliefs. *Professional Psychology: Research and Practice* 30(1):45–50. See also G. A. L. Mazzoni, E. F. Loftus, A. Seitz, and S. J. Lynn. 1999. Changing beliefs and memories through dream interpretation. *Applied Cognitive Psychology* 13:125–144.

17. A. Miller (ed.). 2004. *The Social Psychology of Good and Evil.* New York: Guilford.

18. M. K. Johnson and C. L. Raye. 2000. Cognitive and brain mechanisms of false memories and beliefs. In D. Schacter and E. Scarry (eds.), *Memory, Brain, and Belief.* Cambridge, MA: Harvard University Press.

19. D. Schacter. 1996. *Searching for Memory: The Brain, the Mind, and the Past.* New York: Basic Books.

20. D. Schacter and E. Scarry (eds.). *Memory, Brain, and Belief.* Cambridge, MA: Harvard University Press, p. 3.

21. J. G. Seamon, C. R. Luo, J. J. Kopecky, et al. 2002. Are false memories more dif-

ficult to forget than accurate memories? The effect of retention interval on recall and recognition. *Memory and Cognition* 30(7):1054–1064.

22. V. Goel and R. J. Dolan. 2003. Explaining modulation of reasoning by belief. *Cognition* 87(1):B11–22.

23. R. McNally. 2003. *Remembering Trauma.* Cambridge, MA: Harvard University Press.

24. G. L. Wells and E. F. Loftus. 2003. Eyewitness memory for people and events. In A. M. Goldstein (ed.), *Handbook of Psychology,* Vol. 11, *Forensic Psychology.* New York: Wiley, pp. 149–160.

25. B. Gonsalves and K. A. Paller. 2000. Neural events that underlie remembering something that never happened. *Nature Neuroscience* 3(12):1316–1321.

26. K. A. Wade, M. Garry, J. D. Read, and D. S. Lindsay. 2002. A picture is worth a thousand lies: Using false photographs to create false childhood memories. *Psychonomic Bulletin and Review* 9(3):597–603.

27. M. Garry and K. A. Wade. 2005. Actually, a picture is worth less than 45 words: Narratives produce more false memories than photographs do. *Psychonomic Bulletin and Review* 12(2):359–366.

28. C. Frith and R. J. Dolan. 2000. The role of memory in the delusions associated with schizophrenia. In D. Schacter and E. Scarry (eds.), *Memory, Brain, and Belief.* Cambridge, MA: Harvard University Press, p. 131.

29. A. Schnider. 2001. Spontaneous confabulation, reality monitoring, and the limbic system—A review. *Brain Research: Brain Research Reviews* 36(2–3):150–160.

30. M. A. Conway, C. W. Pleydell-Pearce, S. E. Whitecross, and H. Sharpe. 2003. Neurophysiological correlates of memory for experienced and imagined events. *Neuropsychologia* 41(3):334–340.

31. C. Huron, C. Servais, and J. M. Danion. 2001. Lorazepam and diazepam impair true, but not false, recognition in healthy volunteers. *Psychopharmacology* (Berlin). 155(2):204–209.

32. E. Pernot-Marino, J. M. Danion, and G. Hedelin. 2004. Relations between emotion and conscious recollection of true and false autobiographical memories: An investigation using lorazepam as a pharmacological tool. *Psychopharmacology* (Berlin) 175(1):60–67.

33. J. D. Bremner, J. H. Krystal, D. S. Charney, and S. M. Southwick. 1996. Neural mechanisms in dissociative amnesia for childhood abuse: Relevance to the current controversy surrounding the "false memory syndrome." *American Journal of Psychiatry* 153(7 Supplement):71–82.

34. D. Schacter (ed.). 1999. *The Cognitive Psychology of False Memories: A Special Issue of the Journal Cognitive Neuropsychology* London: Taylor and Francis.

35. D. L. Krebs. 2000. The evolution of moral dispositions in the human species. *Annals of the New York Academy of Science* 907:132–148.

36. G. F. Azzone. 2003. The dual biological identity of human beings and the naturalization of morality. *History and Philosophy of the Life Sciences* 25(2):211–241.

37. H. T. Epstein. 2001. An outline of the role of brain in human cognitive development. *Brain and Cognition* 45(1):44–51.

38. C. J. Boyatzis. 2005. Religious and spiritual development in childhood. In R. F. Paloutzian and C. L. Park (eds.), *Handbook of the Psychology of Religion and Spirituality.* New York: Guilford.

39. A. B. Heilbrun, Jr., and M. Georges. 1990. The measurement of principled morality by the Kohlberg Moral Dilemma Questionnaire. *Journal of Personality Assessment* 55(1–2):183–194. See also B. J. Peens and D. A. Louw. 2000. Children's rights: Reasoning and their level of moral development—An empirical investigation. *Medical Law* 19(3):591–612. And see J. Boom, D. Brugman, and P. G. van der Heijden. 2001. Hierarchical structure of moral stages assessed by a sorting task. *Child Development* 72(2):535–548.

40. J. Fowler. 1995. *Stages of Faith: The Psychology of Human Development.* San Francisco, CA: Harper San Francisco. (First published in 1981.)

41. It has been shown that the brain function pattern changes throughout the first year of life, with initial increases in the sensorimotor cortex, thalami, brain stem, and cerebellar vermis. See H. T. Chugani and M. E. Phelps. 1986. Maturational changes in cerebral function in infants determined by [18]FDG positron emission tomography. *Science* 231:840–843. See also H. T. Chugani, M. E. Phelps, and J. C. Mazziotta. 1987. Positron emission tomography study of human brain functional development. *Annals of Neurology* 22:487–497. These are central systems that subserve brain stem reflexes and visuomotor integrative performance that are typically displayed in infant behavior. See H. T. Chugani 1992. Functional brain imaging in pediatrics. *Pediatric Clinics of North America* 39:777–799. However, there are no significant higher cortical functions, and consequently there is no strong evidence of well-integrated cognitive functioning.

42. C. Kennedy and L. Sokoloff. 1957. An adaptation of the nitrous oxide method to the study of the cerebral circulation in children; normal values for cerebral blood flow and cerebral metabolic rate in childhood. *Journal of Clinical Investigation* 36: 1130. See also H. T. Chugani. 1992. Functional brain imaging in pediatrics. *Pediatric Clinics of North America* 39: 777–799.

43. F. Cirulli, A. Berry, and E. Alleva. 2003. Early disruption of the mother-infant relationship: Effects on brain plasticity and implications for psychopathology. *Neuroscience and Biobehavioral Reviews* 27(1–2):73–82.

44. F. Cirulli, E. Alleva, A. Antonelli, and L. Aloe. 2000. NGF expression in the developing rat brain: Effects of maternal separation. *Brain Research: Developmental Brain Research* 123(2):129–134. See also C. M. Kuhn and S. M. Schanberg. 1998. Responses to maternal separation: Mechanisms and mediators. *International Journal of Developmental Neuroscience* 16:261–270.

45. M. R. Gunnar. 1998. Quality of early care and buffering of neuroendocrine stress reactions: Potential effects on the developing human brain. *Preventive Medicine* 27(2):208–211. See also J. E. Black. 1998. How a child builds its brain: Some lessons from animal studies of neural plasticity. *Preventive Medicine* 27, 168–171.

46. M. S. Gazzaniga (ed.). 2000. *The New Cognitive Neurosciences.* Cambridge, MA: MIT Press.

47. D. D. Francis, J. Diorio, P. M. Plotsky, and M. J. Meaney. 2002. Environmental enrichment reverses the effects of maternal separation on stress reactivity. *Journal of Neuroscience* 22(18):7840–7843.

48. As visuospatial and visuosensorimotor integrative functions are acquired and primitive reflexes are reorganized, there is increasing activity in the primary visual cortex, parietal and temporal regions, basal ganglia, and cerebellar hemispheres. See C. Y. André-Thomas and S. A. Dargassies. 1960. The neurological examination of the infant. In *Medical Advisory Committee of the National Spastics Society.* London. See

also A. H. Parmelee and M. D. Sigman. 1983. Perinatal brain development and behavior. In M. Haith and J. Campos (eds.), *Biology and Infancy,* Vol. 2. New York: Wiley. This also coincides with maturation of the EEG at around two to three months of age. See P. Kellaway. 1979. An orderly approach to visual analysis: Parameters of the normal EEG in adults and children. In D. W. Klass and D. D. Daly (eds.), *Current Practice of Clinical Electroencephalography.* New York: Raven.

49. J. M. Allman, A. Hakeem, and K. Watson. 2002. Two phylogenetic specializations in the human brain. *Neuroscientist* 8(4):335–346.

50. Society for Neuroscience Thirty-Third Annual Meeting, New Orleans, 2003.

51. D. A. Trauner, R. Nass, and A. Ballantyne. 2001. Behavioural profiles of children and adolescents after pre- or perinatal unilateral brain damage. *Brain* 124(Part 5):995–1002.

52. P. R. Huttenlocher and C. deCourten. 1987. The development of synapses in striate cortex of man. *Human Neurobiology* 6:1–9. See also P. R. Huttenlocher. 1979. Synaptic density in human frontal cortex—Developmental changes and effects of aging. *Brain Research* 163:195–205.

53. R. Joseph, R. E. Gallagher, W. Holloway, and J. Kahn. 1984. Two brains, one child: Interhemispheric information transfer deficits and confabulatory responding in children aged 4, 7, 10. *Cortex* 20(3):317–331.

54. D. S. O'Leary. 1980. A developmental study of interhemispheric transfer in children aged five to ten. *Child Development* 51(3):743–750.

55. C. J. Boyatzis. 2005. Religious and spiritual development in childhood. In R. F. Paloutzian and C. L. Park (eds.), *Handbook of the Psychology of Religion and Spirituality.* New York: Guilford.

56. Translated for this volume by Mark Robert Waldman.

57. J. L. Barrett, R. A. Richert, and A. Driesenga. 2001. God's beliefs versus mother's: The development of nonhuman agent concepts. *Child Development* 72(1): 50–65.

58. Boyatzis. Religious and spiritual development in childhood, op. cit.

59. M. Finn and J. Gartner (eds.). 1992. *Object Relations Theory and Religion.* New York: Praeger.

60. A-M. Rizzuto. 1979 *The Birth of the Living God: A Psychoanalytic Study.* Chicago, IL: University of Chicago Press.

61. B. Altemeyer and B. Hunsberger. 2005. Fundamentalism and authoritarianism. In R. F. Paloutzian and C. L. Park (eds.), *Handbook of the Psychology of Religion and Spirituality.* New York: Guilford.

62. M. McCullough, B. Giacomo, and L. M. Root. 2005. Religion and forgiveness. In R. F. Paloutzian and C. L. Park (eds.), *Handbook of the Psychology of Religion and Spirituality.* New York: Guilford.

63. R. Coles. 1990. *The Spiritual Life of Children.* Boston, MA: Houghton Mifflin.

64. H. T. Chugani, M. E. Phelps, and J. C. Mazziotta. 1989. Metabolic assessment of functional maturation and neuronal plasticity in the human brain. In C. von Euler, H. Forssberg, and H. Lagercrantz (eds.), *Neurobiology of Early Infant Behavior.* New York: Stockton.

65. Ibid.

66. In one study, researchers found that two-thirds of high school graduates in industrialized countries failed to reach the level of cognitive maturity described in Piaget's stage of adolescent development. See D. Kuhn, J. Langer, L. Kohlberg, and

N. S. Haan. 1977. The development of formal operations in logical and moral judgment. In *Genetic Psychology Monographs* 95:97–188.

67. R. J. Davidson, D. A. Lewis, L. B. Alloy, et al. 2002. Neural and behavioral substrates of mood and mood regulation. *Biological Psychiatry* 52:478–502.

68. M. Banschick. 1992. God representations in adolescence. In M. Finn and J. Gartner (eds.), *Object Relations Theory and Religion.* New York: Praeger.

69. P. L. Benson, M. J. Donahue, and J. A. Erickson. 1989. Adolescence and religion: A review of the literature from 1970 to 1986. *Research in the Social Scientific Study of Religion* 1:153–181. See also V. King, G. H. Elder, and L. B. Whitbeck. 1997. Religious involvement among rural youth: An ecological and life-course perspective. *Journal of Research on Adolescence* 7:431–456.

70. C. Smith, M. L. Denton, R. Faris, and M. Regnerus. 2002. Mapping American adolescent religious participation. *Journal for the Scientific Study of Religion* 13: 175–195.

71. C. A. Markstrom. 1999. Religious involvement and adolescent psychosocial development. *Journal of Adolescence* 22(2):205–221.

72. M. R. Levenson, C. M. Aldwin, and M. D'Mello. 2005. Religious development from adolescence to middle adulthood. In R. F. Paloutzian and C. L. Park (eds.), *Handbook of the Psychology of Religion and Spirituality.* New York: Guilford.

73. L. Miller, V. Warner, P. Wickramaratne, and M. Weissman. 1997. Religiosity and depression: Ten-year follow-up of depressed mothers and offspring. *Journal of the American Academy of Child and Adolescent Psychiatry* 36(10):1416–1425.

74. L. Miller and B. S. Kelley. 2005. Relationships of religiosity and spirituality with mental health and psychopathology. In R. F. Paloutzian and C. L. Park (eds.), *Handbook of the Psychology of Religion and Spirituality.* New York: Guilford.

75. A. B. Newberg and A. Alavi. 1997. Neuroimaging in the in vivo measurement of regional function in the aging brain. In S. U. Dani, A. Hori, and G. F. Walter (eds.), *Principles of Neural Aging.* Amsterdam: Elsevier Science.

76. W. Crain. 1985. *Theories of Development.* Englewood Cliffs, NJ: Prentice-Hall.

77. Darwin was seventy-two years old when he wrote his autobiography. Originally written for his family, it was first published in 1887 and edited by his son, Francis Darwin, under the title *The Life and Letters of Charles Darwin.* Later editions removed various religious comments Darwin had made. These weren't restored until 1958 with the publication of Nora Barlow's edition, *The Autobiography of Charles Darwin, 1809–1882* (London: Collins).

78. H. Gardner. 1983. *Frames of Mind: The Theory of Multiple Intelligences.* New York: Basic Books.

79. M. R. Levenson, C. M. Aldwin, and M. D'Mello. 2005. Religious development from adolescence to middle adulthood. In R. F. Paloutzian and C. L. Park (eds.), *Handbook of the Psychology of Religion and Spirituality.* New York: Guilford.

80. M. W. Pratt, R. Diessner, A. Pratt, et al. 1996. Moral and social reasoning and perspective taking in later life: A longitudinal study. *Psychology of Aging* 11(1):66–73.

Chapter 6: Ordinary Criminals Like You and Me (Pages 132–164)

1. C. Turnbull. 1961. *The Forest People.* New York: Simon and Schuster.

2. C. Turnbull. 1972. *The Mountain People.* New York: Simon and Schuster.

3. As reported in July 2004 in the online newsletter of The Family, an international volunteer Christian fellowship: http://www.thefamily.org/work/article.php3?id=795.

4. Richard Hoffman's website, http://home1.gte.net/hoffmanr, contains commentaries and photographs of the tribe.

5. The Hammurabi stele was discovered in 1901 and is now kept in the Louvre. Numerous descriptions of this ancient code can be found in books and on various academic websites, including www.kchanson.com.

6. V. Csanyi. 2000. *If Dogs Could Talk.* New York: North Point.

7. The following are two excellent books on this topic. R. Sapolsky. 2004. *Why Zebras Don't Get Ulcers,* 3rd ed. New York: Owl; and D. Goleman. 2003. *Destructive Emotions.* New York: Bantam.

8. Goleman, *Destructive Emotions,* op. cit.

9. J. S. Lerner, L. Z. Tiedens, and R. M. Gonzalez. 2005. Toward a model of emotion-specific influences on judgment and decision-making: Portrait of the angry decision maker. (Manuscript under review, Carnegie Mellon University.)

10. C. V. Caldicott and K. Faber-Langendoen. 2005. Deception, discrimination, and fear of reprisal: Lessons in ethics from third-year medical students. *Academic Medicine* 80(9):866–873.

11. S. Nichols and R. Mallon. 2005. Moral dilemmas and moral rules. *Cognition.* (Published electronically in advance of print.)

12. National Institutes of Health. 2001. Teenage brain: A work in progress—A brief overview of research into brain development during adolescence. Publication No. 01-4929.

13. J. Moll, R. de Oliveira-Souza, I. E. Bramati, and J. Grafman. 2002. Functional networks in emotional moral and nonmoral social judgments. *Neuroimage* 16(3 Part 1):696–703.

14. K. N. Ochsner, R. D. Ray, J. C. Cooper, et al. 2004. For better or for worse: Neural systems supporting the cognitive down-and up-regulation of negative emotion. *Neuroimage* 23(2):483–499.

15. R. Adolphs. 2003. Cognitive neuroscience of human social behaviour. *Nature Reviews: Neuroscience* 4(3):165–178.

16. R. J. R. Blair and L. Cipolotti. 2000. Impaired social response reversal: A case of acquired sociopathy. *Brain* 123:1122–1141.

17. A. Damasio. 1994. *Descartes' Error: Emotion, Reason, and the Human Brain.* New York: Avon Books.

18. Helmuth, L. 2001. Cognitive neuroscience. Moral reasoning relies on emotion. *Science* 293(5537):1971–2.

19. Goleman, *Destructive Emotions,* op. cit.

20. D. M. Fessler, A. P. Arguello, J. M. Mekdara, and R. Macias. Disgust sensitivity and meat consumption: A test of an emotivist account of moral vegetarianism. *Appetite* 41(1):31–41.

21. C. L. Perry, M. T. McGuire, D. Neumark-Sztainer, and M. Story. 2001. Characteristics of vegetarian adolescents in a multiethnic urban population. *Journal of Adolescent Health* 29(6):406–416.

22. S. A. Klopp, C. J. Heiss, and H. S. Smith. 2003. Self-reported vegetarianism may be a marker for college women at risk for disordered eating. *Journal of the American Dietetic Association* 103(6):745–747.

23. M. Bas, E. Karabudak, and G. Kiziltan. 2005. Vegetarianism and eating disor-

ders: Association between eating attitudes and other psychological factors among Turkish adolescents. *Appetite* 44(3):309–315.

24. H. Takahashi, N. Yahata, M. Koeda, et al. 2004. Brain activation associated with evaluative processes of guilt and embarrassment: An fMRI study. *Neuroimage* 23(3):967–974.

25. J. Tangney and J. Stuewig. 2004. A moral-emotional perspective on evil persons and evil deeds. In A. Miller (ed.), *The Social Psychology of Good and Evil.* New York: Guilford.

26. E. Harmon-Jones, K. Vaughn, S. Mohr, et al. 2004. The effects of empathy on anger-related left frontal cortical activity and hostile attitudes. (Unpublished manuscript, University of Wisconsin, Madison.) Reported in C. Batson, N. Ahmad, and E. Stocks. Benefits and liabilities of empathy-induced altruism. In A. Miller (ed.), *The Social Psychology of Good and Evil.* New York: Guilford.

27. For a comprehensive summary and references to these studies, see C. Batson, N. Ahmad, and E. Stocks. 2004. Benefits and liabilities of empathy-induced altruism. In A. Miller (Ed.). *The Social Psychology of Good and Evil.* New York: Guilford Press.

28. L. Carr, M. Iacoboni, M. C. Dubeau, et al. 2003. Neural mechanisms of empathy in humans: A relay from neural systems for imitation to limbic areas. *Proceedings of the National Academy of Science* 100(9):5497–5502.

29. S. E. Asch. 1956. Studies of independence and conformity: A minority of one against a unanimous majority. *Psychological Monographs* 70.

30. C. Nemeth. 1995. Dissent as driving cognition, attitudes, and judgments. *Social Cognition* 13:273–291.

31. H. Bloom. 2000. *Global Brain: The Evolution of Mass Mind from the Big Bang to the 21st Century.* New York: Wiley.

32. S. Milgram. 1983. *Obedience to Authority.* New York: Harper Perennial.

33. C. Browning. 1992. *Ordinary Men: Reserve Police Battalion 101 and the Final Solution in Poland.* New York: HarperCollins.

34. Department of Health and Human Services. 2003. *Child Maltreatment 2001, Administration on Children, Youth, and Families.* Washington, DC: DHHS.

35. P. Zimbardo. A situationist perspective on the psychology of evil. 2004. In A. Miller (ed.), *The Social Psychology of Good and Evil.* New York: Guilford.

36. For an overview of strategies that have been used to demonstrate how moral behavior can be corrupted, see A. Miller (ed.). 2004. *The Social Psychology of Good and Evil.* New York: Guilford.

37. K. Kirby. 2001. How much access should the media have? *Communicator.* (Published online by the Radio-Television News Directors Association.)

38. J. D. Greene and J. Haidt. 2002. How (and where) does moral judgment work? *Trends in Cognitive Science* 6, 517–523.

39. J. D. Greene, R. B. Sommerville, L. E. Nystrom, et al. 2001. An fMRI investigation of emotional engagement in moral judgment. *Science* 293(5537):2105–2108. See also J. D. Greene, L. E. Nystrom, A. D. Engell, et al. 2004. The neural bases of cognitive conflict and control in moral judgment. *Neuron* 44(2):389–400.

40. B. Goodman. 2003. New brain research helps explain how we do—and don't—reason. *Princeton Alumni Weekly,* January 29.

41. *Hastings Center Magazine,* December 1978.

42. C. D. Batson, E. R. Thompson, G. Seuferling, et al. 1999. Moral hypocrisy: Ap-

pearing moral to oneself without being so. *Journal of Personality and Social Psychology* 77(3):525–537.

43. J. B. Rowe, I. Toni, O. Josephs, et al. 2000. The prefrontal cortex: Response selection or maintenance within working memory? *Science* 288(5471):1656–1660. See also F. Hyder, E. A. Phelps, C. J. Wiggins, et al. 1997. "Willed action": A functional MRI study of the human prefrontal cortex during a sensorimotor task. *Proceedings of the National Academy of Sciences of the United States of America* 94(13):6989–6994. And see C. D. Frith, K. Friston, P. F. Liddle, and R. S. Frackowiak. 1991. Willed action and the prefrontal cortex in man: A study with PET. *Proceedings in Biological Science* 244(1311):241–246.

44. J. L. Price. 2005. Free will versus survival: Brain systems that underlie intrinsic constraints on behavior. *Journal of Comparative Neurology* 493(1):132–139.

45. B. Libet. 1991. Conscious versus neural time. *Nature* 352:27–28.

46. J. L. Muller, M. Sommer, V. Wagner, et al. 2003. Abnormalities in emotion processing within cortical and subcortical regions in criminal psychopaths: Evidence from a functional magnetic resonance imaging study using pictures with emotional content. *Biological Psychiatry* 54(2):152–162.

47. K. A. Kiehl, A. M. Smith, A. Mendrek, et al. 2004. Temporal lobe abnormalities in semantic processing by criminal psychopaths as revealed by functional magnetic resonance imaging. *Psychiatry Research* 130(3):297–312. See also K. A. Kiehl, A. M. Smith, R. D. Hare, et al. 2001. Limbic abnormalities in affective processing by criminal psychopaths as revealed by functional magnetic resonance imaging. *Biological Psychiatry* 50(9):677–684.

48. N. Birbaumer, R. Veit, M. Lotze, et al. 2005. Deficient fear conditioning in psychopathy: A functional magnetic resonance imaging study. *Archives of General Psychiatry* 62(7):799–805.

49. D. J. Stein. 2000. The neurobiology of evil: Psychiatric perspectives on perpetrators. *Ethnography and Health* 5(3–4):303–315.

50. HBO Home Video. 2004. *The Iceman Interviews.*

51. Kiehl, Smith, Mendrek, et al. Temporal lobe abnormalities, op. cit.

52. D. S. Kosson, Y. Suchy, A. R. Mayer, and J. Libby. 2002. Facial affect recognition in criminal psychopaths. *Emotion* 2(4):398–411.

53. P. Nakaji, H. S. Meltzer, S. A. Singel, and J. F. Alksne. 2003. Improvement of aggressive and antisocial behavior after resection of temporal lobe tumors. *Pediatrics* 112(5):e430.

54. R. Z. Goldstein, and N. D. Volkow. 2002. Drug addiction and its underlying neurobiological basis: Neuroimaging evidence for the involvement of the frontal cortex. *American Journal of Psychiatry* 159(10):1642–1652.

55. R. C. Gur. 2005. Brain maturation and its relevance to understanding criminal culpability of juveniles. *Current Psychiatry Reports* 7(4):292–296.

56. Sapolsky, *Why Zebras Don't Get Ulcers,* op. cit.

Chapter 7: Nuns, Buddhists, and the Reality of Spiritual Beliefs (Pages 167–190)

1. R. Bucke. 1879. *Man's Moral Nature.* New York, Putnam's. Reported in W. James. 1890. *The Principles of Psychology.* Henry Holt.

2. R. Bucke. *Cosmic Consciousness: A Study in the Evolution of the Human Mind.* New York: Dutton, 1968. (Originally published 1901.)

3. The photograph of Richard Bucke reproduced here is in public domain.

4. A. Newberg, A. Alavi, M. Baime, et al. 2001. The measurement of regional cerebral blood flow during the complex cognitive task of meditation: A preliminary SPECT study. *Psychiatry Research: Neuroimaging* 106(2):113–122.

5. Bucke, *Cosmic Consciousness,* op. cit.

6. The author of this Christian mystical text is unknown, although it has been attributed to an English cloistered monk of the time. The excerpt here is from an adaptation: Evelyn Underwood. 1922. *A Book of Contemplation the Which Is Called the Cloud of Unknowing, in Which a Soul Is One with God,* ed. John M. Watkins from a manuscript in the British Museum, and published by John M. Watkins.

7. T. Keating. 1994. *Intimacy with God.* New York: Crossroad.

8. "The Dalai Lama." 1997. *Mother Jones,* November-December. Interviewed by Robert Thurman.

9. A. Newberg, M. Pourdehnad, A. Alavi, and E. G. d'Aquili. 2003. Cerebral blood flow during meditative prayer: Preliminary findings and methodological issues. *Perceptual and Motor Skills* 97(2):625–630.

10. Y. Joanette, P. Goulet, and D. Hannequin. 1990. *Right Hemisphere and Verbal Communication.* New York: Springer-Verlag.

11. D. H. Ingvar. 1994. The will of the brain: Cerebral correlates of willful acts. *Journal of Theoretical Biology* 171:7–12. See also the following. C. D. Frith, K. Friston, P. F. Liddle, and R. S. Frackowiak. 1991. Willed action and the prefrontal cortex in man: A study with PET. *Proceedings in Biological Science* 244(1311):241–246. M. I. Posner and S. E. Petersen. 1990. The attention system of the human brain. *Annual Review of Neuroscience* 13:25–42. J. V. Pardo, P. T. Fox, and M. E. Raichle. 1991. Localization of a human system for sustained attention by positron emission tomography. *Nature* 349:61–64.

12. K. Vogeley, P. Bussfeld, A. Newen, et al. 2001. Mind reading: Neural mechanisms of theory of mind and self-perspective. *NeuroImage* 14:170–181.

13. A. Damasio. 1994. *Descartes' Error: Emotion, Reason, and the Human Brain.* New York: Avon.

14. O. Muramoto. 2004. The role of the medial prefrontal cortex in human religious activity. *Medical Hypotheses* 62(4):479–485. See also N. P. Azari, J. Nickel, G. Wunderlich, et al. 2001. Neural correlates of religious experience. *European Journal of Neuroscience* 13(8):1649–1652.

15. D. J. Bucci, M. Conley, and M. Gallagher. 1999. Thalamic and basal forebrain cholinergic connections of the rat posterior parietal cortex. *Neuroreport* 10:941–945.

16. It should also be clearly stated that many different parts of the brain are associated with different types of religious and spiritual experiences. In recent years, a number of researchers have focused on specific areas of the brain as a "God spot." In 1997, Dr. Vilayanur Ramachandran, a neuroscientist at the University of California, San Diego, presented a paper at the Annual Conference of the Society of Neuroscience (October 1997, abstract #519.1, vol. 23, Society of Neuroscience) entitled "The neural basis of religious experience" and argued for the temporal lobes being a prime mover in this regard. Michael Persinger also focused on the temporal lobes in his research on sensed presences, and Matthew Alper, author of *The "God" Part of the Brain* (Rogue

Press, 1998) further argued for a particular part of the brain that is involved in religious experience. However, my research with numerous religious and nonreligious practitioners strongly suggests that there is no God "part" or "module," but rather a complex network involving virtually the entire brain when these rich and diverse experiences are elicited. We can point to specific areas of the brain that may be associated with specific components of religious experiences, but since there are numerous ways to perceive, think about, or meditate upon God, each method of meditation or prayer will affect the brain's function in slightly different ways.

17. J. L. Armony and J. E. LeDoux. 2000. How danger is encoded. In M. S. Gazzaniga (ed.), *The New Cognitive Neurosciences.* Cambridge, MA: MIT Press, pp. 1073–1074.

18. P. Maquet, J. Peters, J. Aerts, et al. 1996. Functional neuroanatomy of human rapid-eye-movement sleep and dreaming. *Nature* 383(6596):163–166.

19. T. W. Kjaer, I. Law, G. Wiltschiotz, et al. 2002. Regional cerebral blood flow during light sleep—a [^{15}O]H_2O-PET study. *Journal of Sleep Research* 11(3):201–207.

20. R. P. Vertes. 2002. Analysis of projections from the medial prefrontal cortex to the thalamus in the rat, with emphasis on nucleus reuniens. *Journal of Comparative Neurology* 442(2):163–187.

21. A. B. Newberg and J. Iversen. 2003. The neural basis of the complex mental task of meditation: Neurotransmitter and neurochemical considerations. *Medical Hypotheses* 61(2):282–291.

22. A. J. McDonald, F. Mascagni, and L. Guo. Projections of the medial and lateral prefrontal cortices to the amygdala: A *Phaseolus vulgaris* leucoagglutinin study in the rat. *Neuroscience* 71(1):55–75.

23. T. W. Kjaer, C. Bertelsen, P. Piccini, et al. 2002. Increased dopamine tone during meditation-induced change of consciousness. *Brain Research: Cognitive Brain Research* 13(2):255–259.

24. S. W. Lazar, G. Bush, R. L. Gollub, et al. 2000. Functional brain mapping of the relaxation response and meditation. *Neuroreport* 11(7):1581–1585.

25. J. L. Saver and J. Rabin. 1997. The neural substrates of religious experience. *Journal of Neuropsychiatry and Clinical Neuroscience* 9(3):498–510.

26. For information concerning research conducted at David McCormick's lab, see http://info.med.yale.edu/neurobio/mccormick. For other research concerning the relationship between consciousness and the thalamus, see the following. C. Vakalopoulos. 2005. A scientific paradigm for consciousness: A theory of premotor relations. *Medical Hypotheses* 65(4):766–784. A. Germain, E. A. Nofzinger, D. J. Kupfer, and D. J. Buysse. 2004. Neurobiology of non-REM sleep in depression: Further evidence for hypofrontality and thalamic dysregulation. *American Journal of Psychiatry* 161(10):1856–1863.

27. T. Bal and D. A. McCormick. 1996. What stops synchronized thalamocortical oscillations? *Neuron* 17(2):297–308. See also M. Nakajima, M. Yasue, N. Kaito, et al. 1991. [A case of visual allesthesia]. *No To Shinkei* 43(11):1081–1085. (In Japanese.)

28. D. Hackleman. 1985. The significance of Ellen White's head injury. *Adventist Currents.* (This essay can be found online at www.ellenwhite.org. Hackleman, as noted in the text, was the editor of *Adventist Currents.*)

29. Ibid.

30. A. Dietrich. 2004. The cognitive neuroscience of creativity. *Psychonomic Bul-*

letin and Review 11(6):1011–1026. P. Gilbert. 2001. An outline of brain function. *Cognitive Brain Research* 12:61–74.

31. Dietrich, The cognitive neuroscience of creativity, op. cit.

32. L. Zhang, R. Zhou, X. Li, et al. 2006. Stress-induced change of mitochondria membrane potential regulated by genomic and non-genomic GR signaling: A possible mechanism for hippocampus atrophy in PTSD. *Medical Hypotheses.* (Electronic publication in advance of print.) See also J. L. Warner-Schmidt and R. S. Duman. 2006. Hippocampal neurogenesis: Opposing effects of stress and antidepressant treatment. *Hippocampus.* (Electronic publication ahead of print.)

33. K. Vogeley, M. Kurthen, P. Falkai, and W. Maier. 1999. Essential functions of the human self model are implemented in the prefrontal cortex. *Consciousness and Cognition* 8(3):343–363.

34. J. S. Lerner, L. Z. Tiedens, and R. M. Gonzalez. 2005. Portrait of the angry decision maker. (Manuscript under review, Carnegie Mellon University.)

35. M. Juergensmeyer. 2000. *Terror in the Mind of God: The Global Rise of Religious Violence.* Berkeley: University of California Press.

Chapter 8: Speaking in Tongues (Pages 191–214)

1. R. Balmer. 2001. *Religion in 20th Century America.* Oxford: Oxford University Press.

2. F. Bartleman. 2000. *Azusa Street.* New Kensington, PA: Whitaker House.

3. A well-documented account of the Pentecostal movement is included in D. K. Bernard. 1995. *A History of Christian Doctrine.* Hazelwood, MO: Word Aflame. Some other accounts, particularly those found on Internet sites and in evangelical pamphlets, include biased and inaccurate material.

4. V. Synan. 1997. *The Holiness-Pentecostal Tradition: Charismatic Movements in the Twentieth Century,* 2nd ed. Grand Rapids, MI: Eerdmans. (Synan is a dean at Regent University, a Christian school in Virginia Beach, Virginia.)

5. According to some evangelical statistics, evangelical and charismatic groups have 500 million members. However, many of those groups are now defunct. Furthermore, few statistics have attempted to identify the different sects within the evangelical movement. The *Princeton Religion Research Report 2002* found that 45 percent of the people polled considered themselves born-again or evangelical, but this figure does not tell us about Pentecostal groups. For example, many Baptists, Mormons, and Catholics say they have been born again, but they are not part of the evangelical movement; nor do they speak in tongues. Finally, one must even be cautious about statistics published by prestigious academic presses. For example, a recent publication by Oxford University Press, *World Christian Encyclopedia,* ed. by D. Barrett et al. (2001), was criticized by *Library Journal* for presenting "utterly confusing statistics, some highly suspect, culturally biased, and anthropologically useless."

6. Barna Research Group. 2001. *Religious Beliefs Vary Widely by Denomination,* June 25. At www.barna.org.

7. J. Smith. 1897. *History of the Church of Jesus Christ of Latter-Day Saints.* Salt Lake City, UT: Deseret.

8. W. Williamson. 1992. *An Encyclopedia of Religions in the United States: 100 Religious Groups Speak for Themselves.* Eau Claire, WI: Crossroad.

9. J. T. Titon. 1978. Some recent Pentecostal revivals: A report in words and photographs. *Georgia Review* 32:579–605. On Titon's website (http://www.cs.indiana.edu/~port/teach/relg/pentacostal.revival.htm), phoneticians Linda Ferrier of Tufts and Bob Port of Indiana University provided the phonetic transcriptions.

10. W. Samarin. 1972. Variation and variables in religious glossolalia. In D. Haymes (ed.), *Language in Society.* Cambridge: Cambridge University Press.

11. Personal communications to the authors.

12. W. Cohn. 1968. Personality, Pentecostalism, and glossolalia: A research note on some unsuccessful research. *Canadian Review of Sociology and Anthropology* 5(1):36–39.

13. K. Livingston. 2005. Religious practice, brain, and belief. *Journal of Cognition and Culture* 5:1–2.

14. N. P. Spanos, W. P. Cross, M. Lepage, and M. Coristine. 1986. Glossolalia as learned behavior: An experimental demonstration. *Journal of Abnormal Psychology* 95(1):21–23.

15. B. Grady and K. M. Loewenthal. 1997. Features associated with speaking in tongues (glossolalia). *British Journal of Medical Psychology* 70:185–191.

16. V. H. Hine. 1969. Pentecostal glossolalia: Toward a functional interpretation. *Journal for the Scientific Study of Religion* 8(2):21. See also A. Lovekin and H. N. Malony. 1977. Religious glossolalia: A longitudinal study of personality changes. *Journal for the Scientific Study of Religion* 16(4):383–393.

17. A. G. Hempel, J. R. Meloy, R. Stern, et al. 2002. Fiery tongues and mystical motivations: Glossolalia in a forensic population is associated with mania and sexual/religious delusions. *Journal of Forensic Science* 47(2):305–312.

18. E. Koic, P. Filakovic, S. Nad, and I. Celic. 2005. Glossolalia. *Collegium Antropologicum* 29(1):373–379.

19. L. Francis and M. Robbins. 2003. Personality and glossolalia: A study among male evangelical clergy. *Pastoral Psychology* 51:5.

20. S. H. Louden and L. J. Francis. 2001. Are Catholic priests in England and Wales attracted to the charismatic movement emotionally less stable? *British Journal of Theological Education* 2:65–76. See also M. Robbins, J. Hair, and L. J. Francis. 1999. Personality and attraction to the charismatic movement: A study among Anglican clergy. *Journal of Beliefs and Values* 20:239–246.

21. Hempel, Meloy, Stern, et al. Fiery tongues and mystical motivations, op. cit.

22. D. M. Wegner. 2003. The mind's best trick: How we experience conscious will. *Trends in Cognitive Science* 7(2):65–69.

23. A. Shen. 2000. Free will hunting: Dan Wegner probes the relation between mind and action. *Harvard University Gazette,* October 26.

24. From the Tabernacle Baptist Church website: http://www.tbaptist.com/aab/tongues.htm.

25. John McGrew's example of glossolalia was taken from the following web address: http://www.psynt.iupui.edu/Users/jmcgrew/B365/spirituality%20and%20health/spirituality%20and%20health%20printable%20file.htm.

26. George A. Boyd runs Mudrashram Institute of Spiritual Studies: http://www.mudrashram.com/howmantraswork.html.

27. E. Koic, P. Filakovic, S. Nad, and I. Celic. 2005. Glossolalia. *Collegium Anthropologicum* 29(1):373–379.

28. M. A. Persinger and P. M. Valliant. 1985. Temporal lobe signs and reports of subjective paranormal experiences in a normal population: A replication. *Perceptual and Motor Skills* 60(3):903–909.

29. J. N. Booth, S. A. Koren, and M. A. Persinger. 2005. Increased feelings of the sensed presence and increased geomagnetic activity at the time of the experience during exposures to transcerebral weak complex magnetic fields. *International Journal of Neuroscience* 115(7):1053–1079.

30. N. Jausovec and K. Habe. 2005. The influence of Mozart's sonata K. 448 on brain activity during the performance of spatial rotation and numerical tasks. *Brain Topography* 17(4):207–218.

31. S. Krippner (ed.). 1972. The plateau experience: A. H. Maslow and others. *Journal of Transpersonal Psychology* 4(2):107–120.

32. W. James. 1902. *The Varieties of Religious Experience.* (A study in human nature, the Gifford Lectures on natural religion delivered at Edinburgh in 1901–1902.) New York: Longmans, Green, and Co.

Chapter 9: The Atheist Who Prayed to God (Pages 215–245)

1. Population statistics concerning atheists and other nonreligious individuals can be found at http://atheistempire.com/main.html, which includes surveys from the Pew Research Center, *Encyclopaedia Brittanica,* and the American Religious Identification Survey sponsored by City University of New York. Another excellent source for religious statistics is www.adherents.com

2. W. Jagodzinski and A. Greeley. 1991. "The Demand for Religion: Hard Core Atheism and 'Supply Side' Theory." A copy of this paper can be found on Greeley's website, www.agreeley.com.

3. P. Zuckerman. 2005. Atheism: Contemporary rates and patterns. To be published in M. Martin (ed.). 2006 (in press). *Cambridge Companion to Atheism.* Oxford: Cambridge University Press. Zuckerman's article can be read at www.pitzer.edu/academics/faculty/zuckerman/atheism.html.

4. Pew Research Center for the People and the Press. 2002. Americans struggle with religion's role at home and abroad, March 20. See http://pewforum.org/publications/reports/poll2002.pdf.

5. Pew Research Center for the People and the Press. 2001. Faith-based funding backed, but church-state doubts abound, April 10.

6. For a partial list, with references, see the following websites: http://atheism.about.com/library/decisions/indexes/bl_1_DecisionIndex.htm and http://religiousfreedom.lib.virginia.edu/court/.

7. There is no generally agreed upon categorization of atheistic views. Different scholars have offered different descriptions, and I have used a variety of sources to outline specific divisions, including G. Bromley (ed.). 1988. *International Bible Encyclopedia.* Grand Rapids: Eerdmans. Militant atheism, for example, has been referred to in various historical books, and is occasionally included as a category by atheist organizations.

8. W. William (ed.). 1992. *An Encyclopedia of Religions in the United States: 100 Religious Groups Speak for Themselves.* New York: Crossroad.

9. For a practical introduction to this American version of Vipassana meditation, see J. Kornfield. 1993. *A Path with Heart.* New York: Bantam.

10. J. D. Kass, R. Friedman, J. Leserman, et al. 1991. Health outcomes and a new index of spiritual experience. *Journal for the Scientific Study of Religion* 30:203–211.

11. Spirituality in Higher Education: A National Study of College Students' Search for Meaning and Purpose. http://www.spirituality.ucla.edu/about/index .html.

12. M. R. Levenson, C. M. Aldwin, and M. D'Mello. 2005. Religious development from adolescence to middle adulthood. In R. F. Paloutzian and C. L. Park (eds.), *Handbook of the Psychology of Religion and Spirituality.* New York: Guilford.

13. W. James. 1902. *The Varieties of Religious Experience.* New York: Longmans, Green, and Co.

14. B. A. Kosmin and S. P. Lachman. 1993. *One Nation under God: Religion in Contemporary American Society.* New York: Harmony. See also B. Kosmin, E. Mayer, and A. Keysar. 2001. *American Religious Identification Survey.* New York: City University of New York Press.

15. O. Haase, W. Schwenk, C. Hermann, and J. M. Muller. 2005. Guided imagery and relaxation in conventional colorectal resections: A randomized, controlled, partially blinded trial. *Diseases of the Colon and Rectum* 48(10):1955–1963. See also the following. L. K. Mannix, R. S. Chandurkar, L. A. Rybicki, et al. 1999. Effect of guided imagery on quality of life for patients with chronic tension-type headache. *Headache* 39(5):326–334. M. B. Thompson and N. M. Coppens. 1994. The effects of guided imagery on anxiety levels and movement of clients undergoing magnetic resonance imaging. *Holistic Nursing Practice* 8(2):59–69. L. Roffe, K. Schmidt, and E. Ernst. 2005. A systematic review of guided imagery as an adjuvant cancer therapy. *Psychooncology* 14(8):607–617.

16. N. G. Waller, B. A. Kojetin, T. J. Bouchard, et al. 1990. Genetic and environmental influences on religious interests, attitudes, and values: A study of twins reared apart and together. *Psychological Science* (1):138–142. See also K. R. Truett, L. J. Eaves, J. M. Meyer, et al. 1992. Religion and education as mediators of attitudes: A multivariate analysis. *Behavior Genetics* 22(1):43–62.

17. D. Hamer. 2004. *The God Gene.* New York: Doubleday.

18. D. E. Comings, N. Gonzales, G. Saucier, et al. 2000. The DRD4 gene and the spiritual transcendence scale of the character temperament index. *Psychiatric Genetics* 10(4):185–189.

19. C. Lorenzi, A. Serretti, L. Mandelli, et al. 2005. 5-HT(1A) polymorphism and self-transcendence in mood disorders. *American Journal of Medical Genetics. Part B, Neuropsychiatric Genetics* 137(1):33–35, August 5.

20. B. M. D'Onofrio, L. J. Eaves, L. Murrelle, et al. 1999. Understanding biological and social influences on religious affiliation, attitudes, and behaviors: A behavior genetic perspective. *Journal of Personality* 67(6):953–984.

21. T. J. Bouchard, Jr., M. McGue, D. Lykken, and A. Tellegen. 1999. Intrinsic and extrinsic religiousness: Genetic and environmental influences and personality correlates. *Twin Research* 2(2):88–98.

22. D. I. Boomsma, E. J. de Geus, G. C. van Baal, and J. R. Koopmans. 1999. A religious upbringing reduces the influence of genetic factors on disinhibition: Evidence for interaction between genotype and environment on personality. *Twin Research* 2(2):115–125.

23. S. W. Lazar, G. Bush, R. L. Gollub, et al. 2000. Functional brain mapping of the relaxation response and meditation. *Neuroreport* 11(7):1581–1585.

24. M. Csikszentmihalyi. 1990. *Flow: The Psychology of Optimal Experience.* New York: Harper and Row.

25. E. Diener and M. E. Seligman. 2002. Very happy people. *Psychological Science* 13(1):81–84.

26. J. R. Nevitt and J. Lundak. 2005. Accuracy of self-reports of alcohol offenders in a rural midwestern county. *Psychological Reports* 96(2):511–514.

27. L. R. Clark, C. Brasseux, D. Richmond, et al. 1997. Are adolescents accurate in self-report of frequencies of sexually transmitted diseases and pregnancies? *Journal of Adolescent Health* 21(2):91–96.

28. P. Zuckerman. 2005. Atheism: Contemporary rates and patterns, op. cit.

29. D. Collins. 2003. Pretesting survey instruments: An overview of cognitive methods. *Quality of Life Research* 12(3):229–238.

30. Barna Group. 1999. Christians are more likely to experience divorce than are non-Christians. December 21. See www.barna.org.

31. American Religious Identification Study (ARIS). 2001. See http://www.gc.cuny.edu/faculty/research_briefs/aris/aris_index.htm.

32. Barna Group. 2002. People's faith flavor influences how they see themselves. August 26. See www.barna.org.

33. J. S. Abramowitz, B. J. Deacon, C. M. Woods, and D. F. Tolin. 2004. Association between Protestant religiosity and obsessive-compulsive symptoms and cognitions. *Depression and Anxiety* 20(2):70–76.

34. L. Guiso, P. Sapienza, and L. Zingales. 2003. People's opium? Religion and economic attitudes. *Journal of Monetary Economics* 50:225–282.

35. Harris Poll No. 11. 2003. *The Religious and Other Beliefs of Americans.* February 26.

36. J. H. Leuba. 1916. *The Belief in God and Immortality: A Psychological, Anthropological, and Statistical Study.* Boston, MA: Sherman, French.

37. J. H. Leuba. 1934. *Harper's Magazine* 169:291–300.

38. E. J. Larson and L. Witham. 1997. *Nature* 386:435–436. See also commentary in *Nature* 394:313. (1998.)

39. E. Ecklund and C. Scheitle. 2005. Religious differences between natural and social scientists: Preliminary results from a study of "Religion among Academic Scientists (RAAS)." (Presented August 14 at the Annual Meetings of the Association for the Sociology of Religion.)

40. S. Hawking. 1988. *A Brief History of Time.* New York: Bantam, 90.

41. H. F. Schaefer. 1994–1995. Stephen Hawking, the big bang, and God, Part 1. *The Real Issue,* November-December 1994. Also Part 2. *The Real Issue,* March/April 1995. (Dr. Schaefer originally presented this lecture at the University of Colorado in the spring of 1994, for a program sponsored by Christian Leadership and other campus ministries. To read this article online, go to www.schaefer.gfmuiuc.org.)

42. N. Barlow (ed.). 1993. *The Autobiography of Charles Darwin 1809–1882.* New York: Norton.

43. This comment was first published in Conference on Science, Philosophy, and Religion in Their Relation to the Democratic Way of Life. 1940. *Science, Philosophy, and Religion: A Symposium.*

44. Ibid.

45. L. R. Gianotti, C. Mohr, D. Pizzagalli, et al. 2001. Associative processing and paranormal belief. *Psychiatry and Clinical Neurosciences* 55(6):595–603.

46. C. Mohr, R. E. Graves, L. R. Gianotti, et al. 2001. Loose but normal: A semantic association study. *Journal of Psycholinguistic Research* 30(5):475–483.

47. R. K. Kurup and P. A. Kurup. 2003. Hypothalamic digoxin, hemispheric chemical dominance, and spirituality. *International Journal of Neuroscience* 113(3):383–393.

48. J. Borg, B. Andree, H. Soderstrom, and L. Farde. 2003. The serotonin system and spiritual experiences. *American Journal of Psychiatry* 160(11):1965–1969.

49. In *Science, Philosophy and Religion, A Symposium,* op. cit.

Chapter 10: Becoming a Better Believer (Pages 246–271)

1. L. Leibovici. 2001. Effects of remote, retroactive intercessory prayer on outcomes in patients with bloodstream infection: Randomised controlled trial. *BMJ* 323(7327):1450–1451.

2. *BMJ* maintains an online discussion board and a list of published responses to this study. Go to http://bmj.bmjjournals.com and search for "Leibovici" or "intercessory prayer." An abstract search of PubMed at http://ncbi.nih.gov will find related abstracts and papers.

3. Andrew M. Thornett, Shehan Hettiaratchy, Carolyn Hemsley, et al. 2002. Effect of retroactive intercessory prayer [letters to the editor]. *BMJ* 324:1037.

4. L. Leibovici. April 29, 2002. *BMJ* 324:1037. Leibovici's response was also posted on *BMJ*'s website.

5. B. Olshansky and L. Dossey. 2003. Retroactive prayer: A preposterous hypothesis? *BMJ* 327:1465–1468.

6. D. Radin. 1997. *The Conscious Universe: The Scientific Truth of Psychic Phenomena.* San Francisco, CA: HarperSanFrancisco.

7. W. Braud. 2000. Wellness implications of retroactive intentional influence: Exploring an outrageous hypothesis. *Alternative Therapies in Health and Medicine* 6(1):37–48.

8. Two excellent books that show how easily statistics can be distorted are: Joel Best. 2001. *Damned Lies and Statistics.* Berkeley: University of California Press. Joel Best. 2004. *More Damned Lies and Statistics.* Berkeley: University of California Press.

9. *Neonatal Circumcision.* 1999. Report 10 of Council on Scientific Affairs (I-99), December, published by American Medical Association. (The article can be read at http://www.ama-assn.org/ama/pub/category/13585.html.)

10. C. Ciesielski-Carlucci, N. Milliken, and N. H. Cohen. 1997. Determinants of decision making for circumcision. *Cambridge Quarterly of Health Ethics* 5:228–236. See also J. D. Tiemstra. 1999. Factors affecting the circumcision decision. *Journal of the American Board of Family Practitioners* 12:16–20.

11. R. F. Palmer, D. Katerndahl, and J. Morgan-Kidd. 2004. A randomized trial of the effects of remote intercessory prayer: Interactions with personal beliefs on problem-specific outcomes and functional status. *Journal of Alternative and Complementary Medicine* 10(3):438–448.

12. J. A. Astin, E. Harkness, and E. Ernst. 2000. The efficacy of "distant healing": A systematic review of randomized trials. *Annals of Internal Medicine* 132(11): 903–910.

13. K. Y. Cha, D. P. Wirth, and R. A. Lobo. 2001. Does prayer influence the success of in vitro fertilization–embryo transfer? *Journal of Reproductive Medicine* 46:781–787.

14. B. Flamm. 2004. The Columbia University 'miracle' study: Flawed and fraud. *Skeptical Inquirer,* September. In part, as a result of the diligent inquiries by Dr. Flamm, the authenticity of this study has been challenged. According to Dr. Flamm: The "lead" author of the study (Lobo) claimed that he did not know about the study until months after it was published; Dr. Cha left the university to run an infertility center in California and refused to comment on the study; and "Dr." Daniel Wirth holds no medical degree; he has a master's degree in parapsychology and a law degree. Wirth has since been sentenced to federal prison for fraud and conspiracy. Columbia University investigated the authors and the study and acknowledged noncompliance with its policies and protocols. Flamm writes, "Specifically, Dr. Lobo never presented the above research to the Institutional Review Board of Columbia-Presbyterian Medical Center."

15. E. Ernst. 2003. Distant healing—An "update" of a systematic review. *Wiener Klinische Wochenschrift* 115(7–8):241–245.

16. Several months prior to the publication of this book, Herbert Benson and associates conducted a well-designed and randomized large-scale study of the effects of intercessory prayer on patients undergoing heart surgery. They found that those who knew they were being prayed for had a "higher incidence of complications." H. Benson, J. A. Dusek, J. B. Sherwood, et al. 2006. Study of the Therapeutic Effects of Intercessory Prayer [STEP] in cardiac bypass patients: A multicenter randomized trial of uncertainty and certainty of receiving intercessory prayer. *American Heart Journal,* 151:934–42. It should be noted that Benson believes that prayer makes an important contribution to our physical health.

17. The online dictionary Wikipedia provides an interesting list of fifty types of biases, with dozens of literature references. These include the following. S. Plous. 1993. *The Psychology of Judgment and Decision Making.* New York: McGraw-Hill. T. Gilovich. 1993. *How We Know What Isn't So: The Fallibility of Human Reason in Everyday Life.* New York: Free Press. T. Gilovich, D. Griffin, and D. Kahneman (eds.). 2002. *Heuristics and Biases: The Psychology of Intuitive Judgment.* Cambridge: Cambridge University Press. A. Miller (Ed.). 2004. *The Social Psychology of Good and Evil.* New York: Guilford.

18. E. Pronin, D. Y. Lin, and L. Ross. 2002. The bias blind spot: Perceptions of bias in self versus others. *Personality and Social Psychology Bulletin* 28:369–381.

19. R. Heuer. 1999. *Psychology of Intelligence Analysis.* Washington, D.C.: U.S. Government, Center for the Study of Intelligence, Central Intelligence Agency. (You can read the entire book, or order it for a nominal fee, by going to http://www.cia.gov/csi/books/19104/.)

20. M. Crichton. 2004. *State of Fear.* New York: HarperCollins.

21. R. Highfield. Embryo cloning cheat resigns in disgrace. www.telegraphco.uk.

22. S. Hawking. 2001. *The Universe in a Nutshell.* New York: Bantam. (The figures I quoted can be found on pages 159 and 165. However, I do not know Hawking's source for the number of books published each year.)

23. E. Hamilton and H. Cairns (eds.). 1961. Plato's apology, in *The Collected Dialogues of Plato.* Princeton, NJ: Bollingen.

24. R. Sapolsky. 2004. *Why Zebras Don't Get Ulcers,* 3rd ed. New York: Owl.

25. Y. I. Sheline, M. H. Gado, and H. C. Kraemer. 2003. Untreated depression and hippocampal volume loss. *American Journal of Psychiatry* 160(8):1516–1518. See also S. Murakami, H. Imbe, Y. Morikawa, et al. 2005. Chronic stress, as well as acute stress, reduces BDNF mRNA expression in the rat hippocampus but less robustly. *Neuroscience Research* 53(2):129–139.

26. R. Joseph. 1990. *Neuropsychology, Neuropsychiatry, and Behavioral Neurology.* New York: Plenum. See also E. R. Kandel and J. H. Schwartz. 1993. *Principles of Neural Science.* New York: Elsevier Science.

27. A. Fugh-Berman. 1997. *Alternative medicine: What works.* Baltimore, MD: Williams and Wilkins. See also B. Q. Hafen, K. J. Karren, K. J. Frandsen, and N. L. Smith 1996. *Mind/Body Health: The Effects of Attitudes, Emotions, and Relationships.* Boston, MA: Allyn and Bacon. And see S. J. Coons, S. L. Sheahan, S. S. Martin, et al. 1994. Predictors of medication noncompliance in a sample of older adults. *Clinical Therapeutics* 16:110–117.

28. A. Ellis. 2001. *Overcoming Destructive Beliefs, Feelings, and Behaviors: New Directions for Rational Emotive Behavior Therapy.* Amherst, NY: Prometheus.

29. D. F. Smith. 2002. Functional salutogenic mechanisms of the brain. *Perspectives in Biology and Medicine* 45(3):319–328.

30. R. J. Davidson, D. A. Lewis, L. B. Alloy, et al. 2002. Neural and behavioral substrates of mood and mood regulation. *Biological Psychiatry* 52(6):478–502.

31. M. Beauregard, J. Levesque, and P. Bourgouin. 2001. Neural correlates of conscious self-regulation of emotion. *Journal of Neuroscience* 21(18):RC165.

32. R. J. Davidson. 2003. Affective neuroscience and psychophysiology: Toward a synthesis. *Psychophysiology* 40(5):655–665.

33. J. Kabat-Zinn. 1982. An outpatient program in behavioral medicine for chronic pain patients based on the practice of mindfulness meditation: Theoretical considerations and preliminary results. *General Hospital Psychiatry* 4:33–47. See also J. Kabat-Zinn, A. O. Massion, J. Kristeller, et al. 1992. Effectiveness of a meditation-based stress reduction program in the treatment of anxiety disorders. *American Journal of Psychiatry* 149(7):936–943.

34. For a comprehensive overview, see A. Harrington. (ed.). 1997. *The Placebo Effect.* Cambridge, MA: Harvard University Press.

35. D. E. Moerman. 2001. Cultural variation in the placebo effect: Ulcers, anxiety, and blood pressure. *Medical Anthropology Quarterly* 14:1, 51–72.

36. V. De Pascalis, C. Chiaradia, and E. Carotenuto. 2002. The contribution of suggestibility and expectation to placebo analgesia phenomenon in an experimental setting. *Pain* 96(3):393–402. See also D. D. Price, L. S. Milling, I. Kirsch, et al. 1999. An analysis of factors that contribute to the magnitude of placebo analgesia in an experimental paradigm. *Pain* 83(2):147–156. And see L. Vase, M. E. Robinson, G. N. Verne, and D. D. Price. 2005. Increased placebo analgesia over time in irritable bowel syndrome (IBS) patients is associated with desire and expectation but not endogenous opioid mechanisms. *Pain* 115(3):338–347.

37. H. Benson and R. Friedman. 1996. Harnessing the power of the placebo effect and renaming it "remembered wellness." *Annual Review of Medicine* 47:193–199.

38. Moerman, Cultural variation in the placebo effect, op. cit.

39. A. J. Barsky, R. Saintfort, M. P. Rogers, and J. F. Borus. 2002. Nonspecific med-

ication side effects and the nocebo phenomenon. *Journal of the American Medical Association* 287(5):622–627.

40. D. B. Morris. 1997. Placebo, pain, and belief: A biocultural model. In A. Harrington (ed.), *The Placebo Effect.* Cambridge, MA: Harvard University Press.

41. T. J. Luparello, H. A. Lyons, E. R. Bleeker, and E. R. McFadden. 1968. Influences of suggestion on airway reactivity in asthmatic subjects. *Psychosomatic Medicine* 30:819–825.

42. A. Shapiro and E. Shapiro. 1997. The placebo: Is it much ado about nothing? In A. Harrington (ed.), *The Placebo Effect.* Cambridge, MA: Harvard University Press.

Acknowledgments

ALL BOOKS REPRESENT A COLLABORATION OF MANY VOICES, and this book is no exception. There are many people we would like to thank—friends, families, colleagues, students, and patients—for they have provided us with the inspiration to share our stories and struggles. Without them, life would feel empty and dull. Mark and I would especially like to thank our agents, Arielle Eckstut and Jim Levine; and our editor, Leslie Meredith, for believing in this project and supporting us through every stage of this book. We also wish to thank all the wonderful people behind the scenes at Free Press who helped to give birth to this book. Finally, I want to thank my colleagues and the staff at the University of Pennsylvania, who have supported my quest for understanding the relationship between spirituality and the human brain.

Since I work in a research environment that requires the aid of many people, I have always written my books and papers from the perspective of "we" as a way of acknowledging everyone's contributions. However, for this book, Mark, my colleague and cowriter, suggested that I use the term "I" since much of the material presented here is based on my neurobiological research of the brain. However, I must emphasize that we have collaborated closely with each other in developing many of the novel concepts presented here. He deserves a great amount of credit, for he has brought many important ideas to this book. Since he won't let me use the word "we," and for the suffering I have endured from his unique brand of humor, I suppose I can repay him by saying that any substantial errors you find in this book can be entirely credited to him.

—ANDY NEWBERG, MD

P.S. It has been a great honor to work with Andy, helping him to bring his research into the public arena. His generosity and open-mindedness have been an inspiration, and his willingness to put up with my humor has gone beyond the call of duty. Thank you, Andy, for being my friend. And to my friend and mentor Jeremy Tarcher, my deepest appreciation. I must also thank Scruffy, my dog, whose unquestioning faith has led me to consider that canines have spiritual beliefs as well. I hope, someday, to be the kind of person he believes I am. In the meantime, if you happen to find any substantial errors in this book, I'm going to blame the dog.

—MARK ROBERT WALDMAN

Index

About the Authors

Andrew Newberg, MD, is an associate professor in the Departments of Radiology and Psychiatry at the Hospital of the University of Pennsylvania and he is an adjunct assistant professor in the Department of Religious Studies. He is also the Director of the Center for Spirituality and the Mind at the University of Pennsylvania. He is Board-certified in internal medicine, nuclear medicine, and nuclear cardiology.

Dr. Newberg has published over one hundred articles, essays and book chapters, and is the co-author of *Why God Won't Go Away: Brain Science and Biology of Belief* (Ballantine, 2001) and *The Mystical Mind: Probing the Biology of Religious Experience* (Augsburg Fortress Publishers, 1999). An overview of his work can be viewed at www.andrewnewberg.com.

Mark Robert Waldman is an associate fellow at the Center for Spirituality and the Mind at the University of Pennsylvania. He is the author of nine books and anthologies on personal relationships, dreamwork, creativity, literature, and writing. He was the founding editor of the academic literature review journal, *Transpersonal Review,* which covered the fields of transpersonal and Jungian psychology, religious studies, and mind/body medicine. He has a counseling practice in Southern California, specializing in relationship dynamics and psychospiritual development. You can contact him at markwaldman@sbcglobal.net, or you can visit his website at www.markrobertwaldman.com.